Change the World
CITY BY CITY

This book is dedicated to all the change makers that are determined to be part of the solution rather than the problem, that take responsibility in crafting a more desirable and sustainable future and that invest themselves to create a better world for all.

Change the World

CITY BY CITY

A Change Maker's Guide To Fast Forward Sustainability

Leen Gorissen

With Erika Meynaerts

The ARTS project was funded by the European Union's 7th Framework Programme for research, technological development and demonstration under the grant agreement no 603654. The creation of this book is a co-investment of VITO and Studio Transitio. All profits generated will be donated to Kom op tegen Kanker, a Flemish non-governmental organisation that strives for healthy living environments and supports the rights of cancer patients to get the best care available.

D/2018/45/279 – ISBN 978 94 014 5357 8 – NUR 758

Cover and interior design: Fulya Toper
Cover photograph: Chris Barbalis
Mechanism icons based on design by Gesine Hildebrandt, Dresden

© VITO & Lannoo Publishers nv, Tielt, 2018.

How to cite this book:
Gorissen, L. & Meynaerts, E. (Eds), with contributions of Barnes, J., Blum, A., Borgstrom, S., Egermann, M., Frantzeskaki, N., Rok, A. and Strenchock, L. 2018. *Change the World City by City. A Change Maker's Guide to Fast Forward Sustainability*. LannooCampus, Leuven, Belgium, ISBN.

LannooCampus Publishers is a subsidiary of Lannoo Publishers,
the book and multimedia division of Lannoo Publishers nv.

All rights reserved.
No part of this publication may be reproduced and/or made public, by means of printing, photocopying, microfilm or any other means, without the prior written permission of the publisher.

LannooCampus Publishers
Erasme Ruelensvest 179 box 101
3001 Leuven
Belgium
www.lannoocampus.be

Contents

Foreword	8
Origins	11
Chapter 1: Introduction	13
Cities in transition	14
Speeding up and scaling positive change	15
About this book	17

PART ONE:
HOW TRANSFORMATIVE INITIATIVES ARE REDESIGNING THEIR CITIES   21

Chapter 2: Change makers, transition initiatives and systems change	23
Chapter 3: Five transformative initiatives in five European cities	27

PART TWO:
HOW CHANGE MAKERS ARE ACCELERATING URBAN SUSTAINABILITY   91

Chapter 4: Mechanisms for acceleration	93
Chapter 5: Overview of sustainability dynamics in five European cities	105

PART THREE: TOWARDS FUTURE-FIT CITIES   **123**

Chapter 6: Key insights   **125**
    Meaningful change is happening organically, everywhere, all the time   127
    Change starts inside   128
    Food is a promoter and unifier for urban sustainability   128
    Sequencing change by working on new patterns   131

Chapter 7: Ten things to know and do to fast forward urban sustainability   **133**
    #1. The next big thing will be a lot of small things   134
    #2. The city of the future is designed for life (not cars)   138
    #3. Human – nature – human nature reconnecting is key   142
    #4. Forget power or status. Urban change makers are motivated by passion and purpose, shifting leadership from ego-based to soul-based   146
    #5. Welcome urban hacking, to create the future is to hack the past   150
    #6. Forget competition. The city of the future is built on collaboration and powered by reciprocal networks   154
    #7. Building a new logic of value creation is the only way out of the status quo   158
    #8. Updating urban social software is foremost about advancing collective learning and enabling collective intelligence and inclusive action   162
    #9. Space is an important resource for innovation. Open up space amid everyday busy-ness in support of business-as-UNusual   166
    #10. Forget certainty, stability and control. The future is made at the sweet spot between order & chaos and unity & diversity   170

Chapter 8: Conclusion   **173**
    Reinventing ourselves as human beings   174
    While change is constant, it is never easy   174

Endings   **176**
Saying thanks   **178**
About the authors   **180**
Endnotes   **181**
Glossary   **183**
Find out more   **195**

> Our struggle for global sustainability
> will be won or lost in cities.
>
> **UN Secretary-General Ban Ki-moon**

Foreword

*By Bart Somers, Mayor of the city of Mechelen, Belgium and
Winner of the World Mayor contest of 2016*

Benjamin Barber passed away on 24 April 2017. On that day, all mayors lost a passionate advocate. After all, in his book *If Mayors Ruled the World*, Benjamin Barber explained the difference between local and national politics in such minute detail. A mayor does not exist to develop abstract ideas or to define ideological dividing lines, but rather to resolve specific problems experienced by real people. In national politics you might be able to score points with walls and trenches, but a politician at local level knows that he must first and foremost build bridges.

This local approach works. Cities are fine examples of 'laboratories' in which solutions are devised for both minor and large-scale issues. Europe – and by extension the rest of the world – is now facing a multitude of challenges. Consider the flows of migrants and the pressure they are putting upon our society, the people's waning trust in traditional institutions, the environmental challenges of climate change and the depletion of natural resources. These are global challenges that must be addressed at local level in the first place.

I have read a great many theories about increased diversity. About how a society composed of different communities, with different origins and religious beliefs, is fuelling tension and unrest. That is certainly true, but as a mayor, I have very little use for such theories. As a mayor, you do not get bogged down in analyses, but instead you work on finding solutions. You need to ensure that those different communities all feel at home in your city, as it is also their city. You can't do that with a single measure and you can't do that alone either. You can only achieve results with the sustained efforts of many. As a mayor, your primary role is that of a coach, to bring people together and to ensure that everyone feels involved. This way, you provide scope for numerous bottom-up initiatives, which – when combined – result in a response at local level to the global migration issue.

The same is true for people's waning trust in the traditional institutions. When you bear responsibility for policy at local level, you know better than anyone that imposing a vision from above does not work. These days, citizens' groups and associations sometimes have more scientific evidence and better technology than officials. Failing to appreciate their expertise will not land well. For this reason, no other level of government experiments more with new forms of participation. For some time now, that input is no longer limited to voting yes or no to a particular proposal. We are now in the stage of co-creation in which citizens develop a proposal themselves by working together. It is inevitable that many of these experiments will find their way to the top level and also influence national and international politics.

A third global challenge that will be solved at local level is that of environmental sustainability and that is what this book is about. The authors highlight that the solutions to the global challenge of climate change and the call for more sustainable policy will have to come from the bottom-up, from our cities. And cities and municipalities are also aware of this. The five examples cited in this book are testament to that. But I also refer to the Covenant of Mayors for Climate and Energy, in which thousands of local administrators are getting involved in order to achieve the abstract European target of reducing CO_2 emissions by 40% by 2030 at local level. It is proof that local policymakers are aware of their duty and responsibility, because we will ultimately have to change our way of living and we will have to produce and consume in a different way. That change will happen at the local level.

By emphasising the importance of cities in the approach to tackling the challenge of climate change, the authors of this book pay tribute to Benjamin Barber. I agree with Barber in my certainty of one thing: if mayors ruled the world, the world itself would be a better place.

Foreword

By Walter Eevers, Research Director of VITO

Important present-day challenges are of a global nature but manifest themselves in ways that are influenced by local conditions. They require an action-oriented and holistic approach of society cutting through and integrating research, technology and innovation. Cities can be characterised by a high concentration of people, finance, knowledge and networks. Because of their density, the increasing amounts of available data and the smarter methods to acquire and use these data, cities have become ideal breeding grounds for testing and implementing disruptive innovations across different societal systems, like mobility, biodiversity, housing etc. Successful innovations, in order to be action-oriented and holistic, require citizen involvement not only as a source of data but more importantly as an adopter of change, a coproducer of solutions. That's what makes the projects described in this book so exciting. They show how involving citizens at a very early stage in the innovation process as well as throughout the development process, can open new solution corridors and accelerate the uptake and upscaling of societal sustainable innovations. This book offers a cornerstone perspective on system change, for our cities, and thereby for a growing majority of people in our world.

Origins

– a personal intro

It all starts with language and stories. Language – as an articulation of reality – is more primordial than strategy, structure or culture. Stories is how we arrange, remember and make sense of knowledge and experiences[1], past and present. What is more, according to cognitive scientist Mark Turner, parables are the roots of the human mind, of our thinking, knowing, acting and creating, and therefore indispensable tools for everyday reasoning. And so it seems appropriate to start this book with a parable. One that shifted my mind.

> It all started when the great forest caught fire. The fire raged and burned. It scared all the animals, small and large. They fled from their homes and gathered at the forest edge. All except one. Instead of being paralysed into inaction, the little hummingbird flew to the river, picked up a droplet of water and flew back into the burning forest to drop the water on the fire. Again she flew to the stream and brought back another drop, and so she continued—back and forth, back and forth. The other animals watched the bold hummingbird brave the enormous fire in terror. They called out to the little hummingbird, warning her of the dangers of the smoke and the heat. They all had opinions and moaned "the fire is too hot", "the smoke is too thick" and "my beak is too small", "my feet will burn"… But the little hummingbird could not be swayed to stop. She persisted in flying to and fro, picking up more water and dropping it, bead by bead, onto the fire. Finally the big bear said, "Little hummingbird, what are you doing?" Without stopping, the little bird looked down at all the animals and said, "I am doing my part."[2]

I am often reminded of this parable because we live in turbulent times today. And more than others, change makers have to deal with a plethora of – often opposite – opinions, viewpoints, recommendations, comments and advice. Usually well-meant but more often than not sprouting from old and reductionist ways of thinking. As the saying 'The only constant is change' portrays, life on Earth is in constant evolution. Sustainability is not an endpoint, but rather a byproduct of regenerative value creation. So the question is not 'Is this initiative big enough and 100% sustainable?' but 'How is this initiative building capacity for evolution?', 'How is this initiative adding value to the larger system?' and 'How is this initiative contributing to sustainability transitions?'. For, an initiative can only be value adding if it is beneficial to the larger system it inhabits and if it can co-evolve with the environment it is embedded in. So, if we want to evolve our systems, we will have to evolve ourselves and our role in the system. And, like Mang and Haggard so eloquently show in their book *Regenerative Development and Design*, it all starts with evolving the way we think about sustainability[3].

This book is about how we can all be the makers of positive change. The future is not something that happens to us, it is the result of what we do in the present. Our individual actions are not able to transform the system in its totality but collectively we are able to create the conditions for systems change. We all have a part to play. We can choose to become part of the solution. The city of the future is not the result of technologies and infrastructure, however smart they may be. The city of the future is what emerges from collective intelligence, collaborative action and unlimited creativity.

And so it is time to tell the story of how urban change makers are scaling beneficial change in their cities. A story that originated in a research project called 'ARTS', about accelerating the transition to sustainability[4]. This book is an attempt to disclose some of the more generic learnings that we gained throughout the project. It focuses on urban transition initiatives and how they contribute to the wider dynamics of change. It shows that urban change makers are important innovators. They are the creators of transition initiatives – the start-ups of a new future – that transform current urban living and thereby shape the pathways to the city of the future.

Leen Gorissen

Chapter 1:
Introduction

The illusion of powerlessness is a greater impediment to change than the power of vested interests.

Flor Avelino, Dutch Research Institute for Transitions

Cities in transition

On Earth Overshoot Day, we have used more resources from nature than our planet can renew in the whole year. In 2016, this was on August the 8th and it illustrates the alarming rate at which humanity extracts resources from the earth. It means that we are overexploiting nature to an extent that we are undermining its life support system by overfishing, overharvesting and deforestation. We simply do not give our Earth a chance to regenerate its sinks and resources – we need more than one planet to fulfil current consumption needs. Earth Overshoot Day highlights the dramatic disconnect that has arisen between humans and nature and is one more evidence point that we are now in the Anthropocene: the era that humans pose an irreversible impact on Earth. Our unstoppable hunger for resources, the ungracious way of heat, beat and treat to make stuff and our exploitative value system are important drivers of the present ecological crises. The sustainability challenges ahead of us, such as climate change, biodiversity loss, resource scarcity and related inequity are so grand and so complex that it may make many people feel powerless.

Yet, new leaders are standing up to resolve these grand challenges by taking responsibility to innovate, affect change and awaken creativity, ingenuity and compassion in their communities. They are civic innovators, urban change makers that shape local responses to global issues. Across the globe, these local responses are proliferating in cities. We call them 'transition initiatives' because they aim to change the way we think, act and organise ourselves to find more sustainable ways of addressing the needs of our society. Transition initiatives showcase that alternative and more sustainable ways of being, living and collaborating together are possible. This way, transition initiatives become pockets of the future in the present. They are the start-ups building the city of tomorrow, they are evolving our social software for the next phase of civilization. One that is more in harmony with the rest of life on the planet.

These transition initiatives are transformative, local answers to our global problems. They are driven by urban change makers from civil society, business, government or their partnerships and insert innovative solutions into their cities. Think of community-supported agriculture initiatives, renewable energy cooperatives, sharing platforms, repair cafés, food forests, urban beekeepers, participatory budgeting approaches etc. Yet, can these typically small-scale interventions impact wider systems' change? Can small alterations within urban communities cascade into big changes on a planetary scale? In other words, are local transition initiatives just operating in the fringe or do they contribute to acceleration of urban sustainability transitions, shifting urban systems to higher levels of sustainability?

In this book, we take you on a journey to five European cities: Brighton, Budapest, Dresden, Genk and Stockholm, to discover how transformative, local initiatives in these cities are shaping a more sustainable future. The findings presented here are based on a collaborative research project funded by the European Commission titled 'Accelerating and Rescaling Transitions to Sustainability' (ARTS project) in which researchers, policymakers, civic innovators, community builders and entrepreneurs came together to create local roadmaps to accelerate progress to sustainability in their cities.

Speeding up and scaling positive change

Cities are hotbeds of consumption and pollution. But cities are also breeding grounds for sustainable innovation and solutions. By becoming more sensitive and responsive to sustainability challenges and concerns, they can become important catalysts to increase the pace of positive change and accelerate the transition to a more benevolent, regenerative way of urban living. So, the questions we ask ourselves while thinking about transformative change in cities are:

- What changes?
- Who is promoting change?
- Is the pace of positive change accelerating and if so, how?

To find answers to these questions, we look at pockets of transformation in cities. To comply with current scientific rigor, we need to set relevant systems boundaries and definitions. We define transition initiatives as 'locally-based activities which aim to drive transformative change of existing societal systems towards higher levels of environmental sustainability' and the system under study is the city region. Environmental sustainability is understood as contributing to conditions of balance, resilience and interconnectedness that allows society to satisfy its needs without exceeding the capacity of its supporting ecosystems to continue to regenerate the services to meet those needs and without diminishing biodiversity. We view these transition initiatives as living and evolving entities that are shaping and are shaped by the contexts they operate in. We thus adopt a multilevel (community, district, city, region, nation) governance approach and specifically set out to understand how transition initiatives, initiated by civil society, business and the government or partnerships of those, help accelerate progress to sustainability in urban settings.

What?
By what changes, we do not focus on the solutions in themselves but rather on the dynamics of change that encompass the way of thinking (culture), doing (practice) and organising (structure). In other words, are transition initiatives fuelling new mindsets, new practices and new arrangements and structures that can be the seedbeds for wider societal change? Widening our understanding from solutions to dynamics of change helps us build a more holistic, systemic understanding of urban transformation.

Who?
The focal unit of analysis are the innovative activities and related actor-networks from the city region, defined as 'transition initiatives', situated within the local governance context, that are focused on driving environmental sustainability. Since sustainability in cities is a collective effort and outcome, we do not only include policies or government interventions but all transformative local actions that improve environmental sustainability locally. We look at initiatives that transition the way we think, act and organise and their 'journeys' in making cities more sustainable by looking at the people driving these solutions and how they interact and synergize.

How?
To study how the pace of positive change can be increased, we looked at aspects of acceleration[5]. From scientific literature, we know that replication, partnering, upscaling, instrumentalising and embedding have been put forward as important mechanisms that promote acceleration[6]. While we theorize that each of these mechanisms can contribute to faster transformation processes, we investigated whether local transition initiatives adopt these mechanisms and whether these can effect wider and faster change. Each of these aspects of acceleration will be explained in more detail in the following chapters and illustrated with findings from the five case cities.

Replicating: The reproduction of alternative ways of thinking, doing and organising from one neighbourhood to another and from one city to others.

Partnering: The formation of partnerships to create synergies and to promote diffusion of alternative ways of thinking, doing and organising within cities.

Upscaling: The increase in the amount of people involved in sustainable alternatives to enable their mainstreaming within cities.

Instrumentalising: The resourcefulness to seize relevant windows of opportunity to support new ways of thinking, doing and organising in cities.

Embedding: The integration of more sustainable ways of thinking, doing and organising in city-regional governance patterns encompassing routines, habits, customs, political and institutional structures.

About this book

This book thus explores the ways transition initiatives are scaling positive and meaningful change in their cities by transitioning urban culture, practice and structure towards higher levels of sustainability. The focus lies on change makers' initiatives and how they contribute to the dynamics of change within their city regions to accelerate progress to sustainability. This approach offers a more exploratory take on sustainability dynamics in cities: How are transition initiatives in cities influencing the spreading of more sustainable ways of thinking, doing and organising in their local context? How are they interrelating with the local government? And how are they creating local and trans-local impact? This way, we offer more insights in how sustainability in cities is evolving. Even though our understanding is work-in-progress, our findings show that cities are places where we can observe sustainability transitions in the making. We expand the current and often technological-myopic understanding of transitions to how people, initiatives and places in our cities play out in transitions. We highlight that place is also important in demarcating transformations: transitions are also about changing places, in confluence of changing ways of doing, thinking, organising. The transition initiatives studied combine these dimensions, and showcase a great variety of sustainable alternatives that work in their place.

The next sections of this book are organised as follows: you can read how transition initiatives are redesigning their cities in part one. Here, we introduce how local transitioneurs can contribute to wider system change, the five cities under study and a diverse set of inspiring local transition initiatives in Brighton, Dresden, Stockholm, Budapest and Genk. In part two, we zoom in on the mechanisms of acceleration and how transition initiatives work to speed up progress towards sustainability in their cities. Each mechanism is explained and illustrated with an example from the cities under study. For readers who want to learn more, we dive deeper into the dynamics of change and how these relate to the mechanisms in the next chapter. Here you will find a lot of examples describing how urban transition initiatives are accelerating meaningful change.

Colour codes and icons will tell you from which city they are and in which domain they are working. In part three we reflect upon the learnings we took from these initiatives on a meta level. In other words, if we zoom out, what generic insights come to light? How do these transition initiatives accelerate sustainability in their city regions? We highlight ten things to know and do to fast forward sustainability in your city. We summarize what this means for transitioning urban systems to sustainability in the conclusions and offer some guidelines for further reading.

Icons for Domains:

Throughout the book, you may encounter words that are not familiar to you. Do not worry, at the end of the book you will find a glossary where we explain these. For changing systems also requires us to change our language and a better understanding of how transitions unfold requires new words to describe the new processes and insights. It is no coincidence that Inuit have so many different words for snow. They use it as an important resource and thus need to specify richer and more precise meanings. The same holds true for transition language. Changing urban culture, practice and structure requires changing our language too. An import feedback that we got from the change makers we worked with was that long, academic and descriptive literature is not appealing to them. That is why we chose to present this book in a format that resembles a city guide that you can browse according to your interests. Bearing the change makers' feedback in mind, we tried to keep it short, focus on what works, highlight insights from change makers and keep references to a bare minimum.

The global crisis of unsustainability is not only a crisis of the hardware of civilization, it is also a crisis of the software of minds. The search for a more sustainable development in the 'developed' world has, so far, been focusing too much on hardware updates, such as new technologies, economic incentives, policies and regulations, and too little on software revisions, that is cultural transformations affecting our ways of knowing, learning, valuing and acting together. The cultural software is, nevertheless, at least as much part of the fundamental infrastructure of a society as its material hardware.

Sacha Kagan, Institute of Sociology and Cultural Organization

PART ONE:

HOW TRANSFORMATIVE INITIATIVES ARE REDESIGNING THEIR CITIES

Chapter 2:

Change makers, transition initiatives and systems change

Between the government and market is a third economic power growing: an entrepreneurial civil society. A fast growing group of initiatives experimenting with new transitional economic models where ecological and social aims are not adjective but in the core of the activity. The purpose is doing well by doing good.

Erwin De Bruyn, Stebo Genk

Transforming cities requires change on all fronts: culture, practice and structure. First, it requires change on the individual level. To change the system we need to change our thinking first. Citizens will have to evolve their mindsets, habits and practices from reinforcing unsustainable systems to developing sustainable alternatives. This entails building ecoliteracy on how everything is interconnected, growing consciousness and courage to change unfitting norms, beliefs and worldviews and cultivating resolution to become part of the solution rather than being part of the problem.

Second, it requires transformation on the structural level. Living healthy, fairly, responsibly and meaningfully on this resource-constraint planet will require us to change the way we organise our value systems and how we organise ourselves. The challenge is to shift our current dominant exploitative value system to a regenerative one – one that is supportive instead of eroding – and to shift outdated organisational formats, optimized for control, demand and efficiency to new ones that are more emancipatory, responsive, adaptive, value adding and creative in times of change[7]. It also requires us to change urban governance in ways that every citizen can become a co-creator and co-investor of a better future. Correspondingly, this means breaking down institutions that are no longer appropriate and replacing them with new ones that are more fit for the future.

Thirdly, it requires us to change the way we do stuff. We need new practices, new tools, new skills, new evaluation criteria, new models to measure progress etc. so that sustainable ways of doing will become ingrained in our every day routines. This means experimenting and improving what we do along the way since there are no blueprints or one-size-fits-all solutions. Rather than looking for silver bullets, systems change is about increasing the capacity for evolution – how we can improve the city with each step that we take so that we can continuously invest in the future rather than lock ourselves into the past.

Urban change makers, regardless of whether they come from research, business, governments or civil society, are instrumental to fuel these evolutionary dynamics. Local change makers are the pioneers that establish transition initiatives, transitioning the way citizens think, act and organise to higher levels of sustainability. In our view, sustainability is much more than reproducing sustainable solutions. It is about going forward and growing our capacity to add rather than extract value from our environment. It is a developmental process of building reciprocal relationships between humans and nature that increases the viability and vitality of both. Mang & Haggard (2016) postulate that "sustainability is actually a byproduct of growing the value that living systems create." So sustainability is not an end-point but a process, a process of value adding that creates conditions beneficial to life[8].

© Sarah Davenport

Change makers in cities are transitioneurs, short for transition entrepreneurs, because their initiatives can be considered as the start-ups of the city of the future. Transitioneurs are found in every sector, from civil society, over policy to business. They introduce new ways of thinking, doing and organising in their city by transferring concepts, ideas, practices on sustainable alternatives to their local context using the expanding array of communication channels, social media and transport opportunities. Transition initiatives intervene in the city *as it is* and showcase urban innovations on how the city *can be*. That way, they contribute to the dynamics of change. Just like start-ups in the business world, they start small by tapping into new ways of value creation. When these interventions create beneficial ripple effects across scales, they can induce system-wide effects. And that is exactly what we set out to study.

Chapter 3:

Five transformative initiatives in five European cities

You never change things by fighting the existing reality.
To change something, build a new model that makes
the existing model obsolete.

Buckminster Fuller

Where are these projects from?

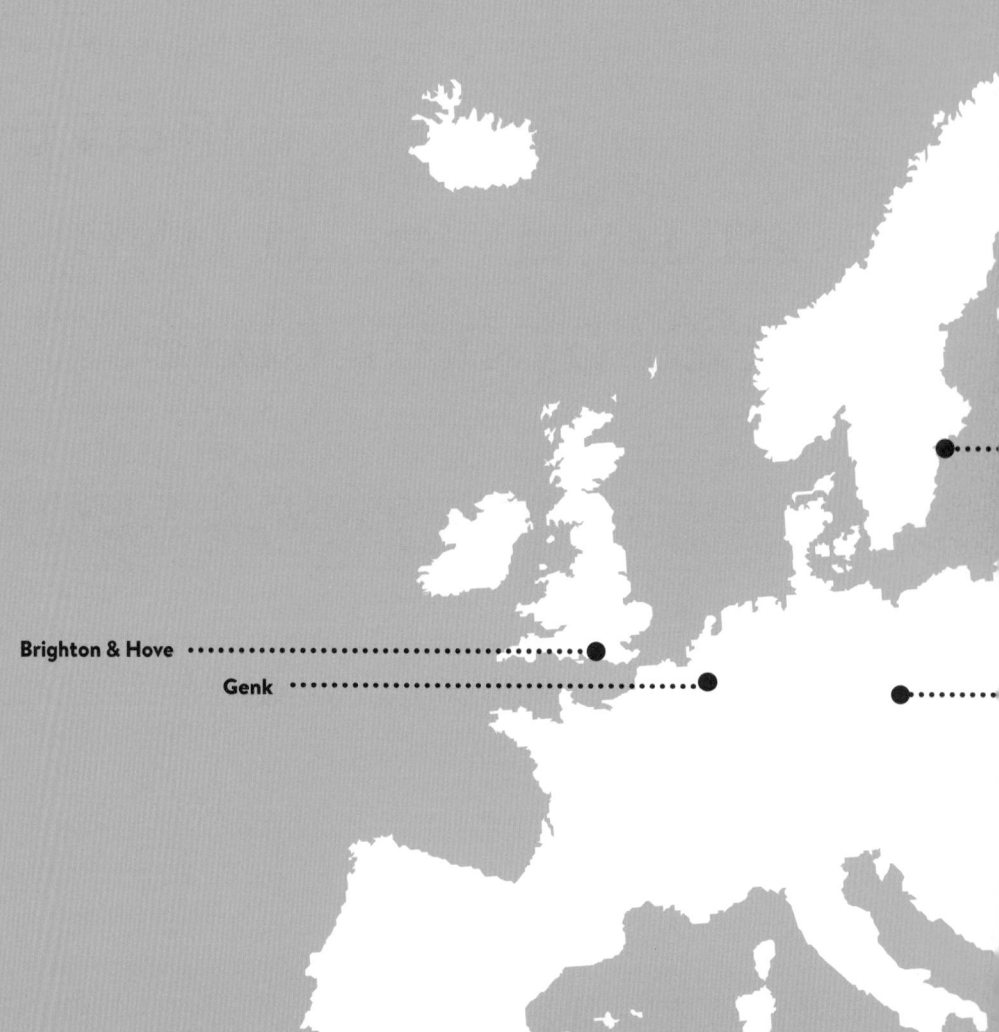

Brighton & Hove

Genk

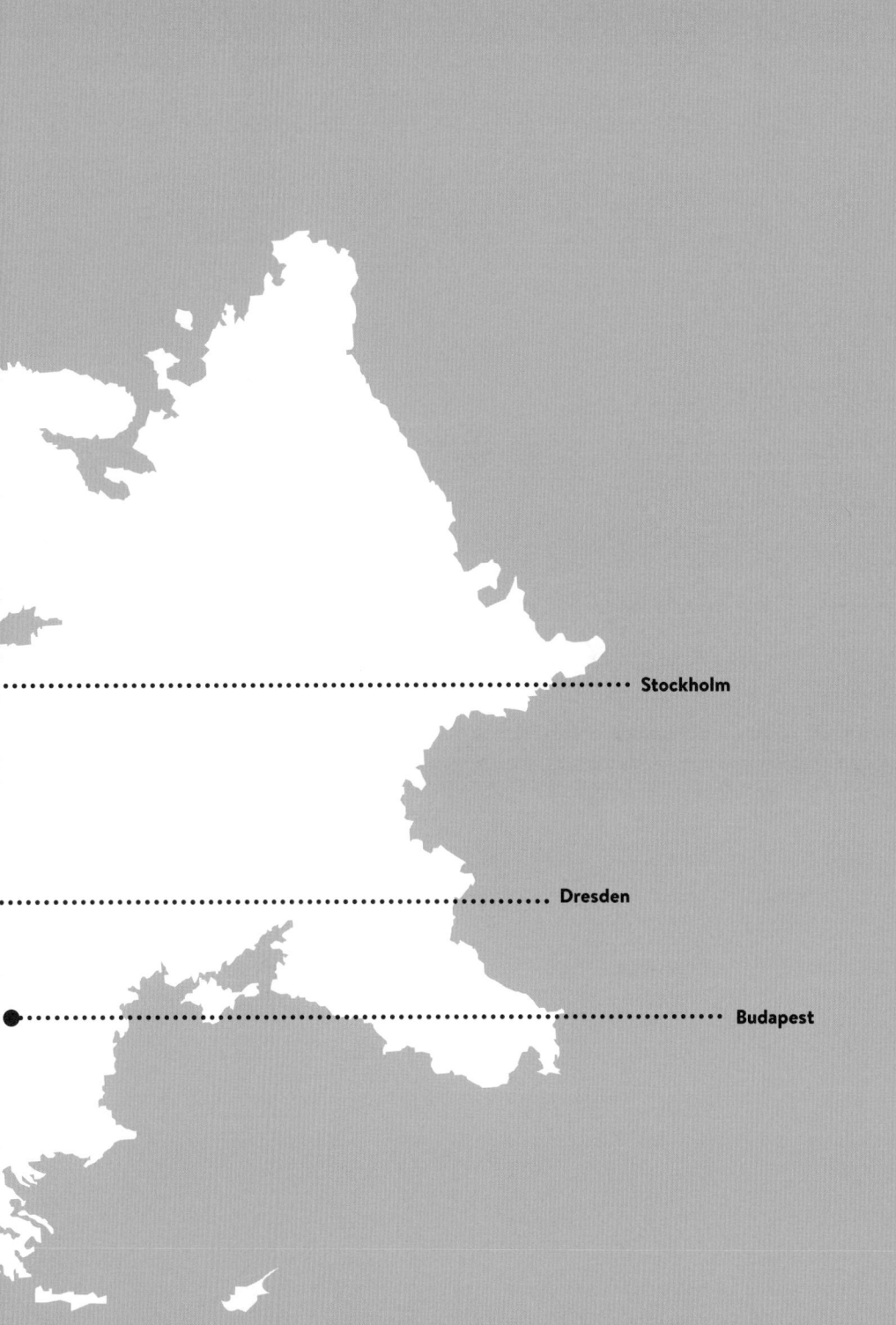

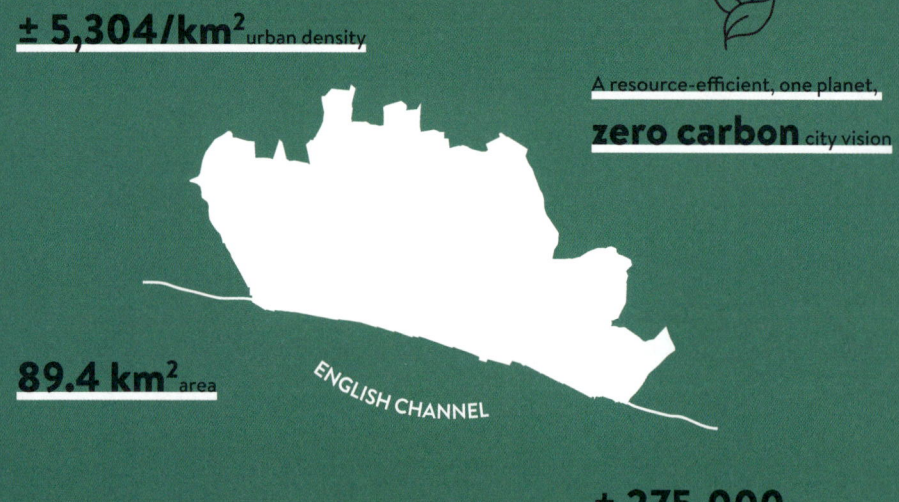

± 5,304/km² urban density

A resource-efficient, one planet, **zero carbon** city vision

89.4 km² area

ENGLISH CHANNEL

± 275,000 inhabitants

Brighton & Hove

Brighton & Hove is a coastal city of approximately 275,000 people located in the South-East of England. It is densely populated (seven times higher than the average for the South East of England) and is forecast to grow by over 5% by 2021. Brighton & Hove is a transport hub and popular tourist destination. The local economy is based predominantly on services, health care, education, and tourism and whilst being situated in one of the most economically developed areas of the UK (the south east), parts of the city also suffer from a high percentage of households in poverty.

Brighton & Hove was the first city in the UK to elect the Green Party into local government in May 2011 (until 2015) and is the only city in the country that is represented by a Green MP in the UK Parliament. From mid 2000s onwards a strong political commitment to sustainability across all parties has anchored sustainability as a priority within local government. In 2013 the city became the world's first designated 'One Planet Living City' through adoption of one planet living principles. Endorsed by the World Wide Fund for Nature the principles set out how the city can live and work within a fair share of the planet's resources.

The city is also pioneering in other ways. It has the highest number of same-sex civil partnerships in the UK, 12% more households not owning a car than the UK average, twice as many people that walk to work than the national average and a whole variety of bottom-up, mainly civil society-led activities promoting more sustainable ways of thinking, doing and organising within the city region. These include communal recycling, cookery projects, coordinated eco-houses open days, bike trains, city 'nature wardens', community gardening, an eco-technology show and sustainability-focused business training programmes. The city also established a City Sustainability Partnership, led by the City Council, to bring together different umbrella bodies and improve coordination across the city region.

Five inspiring local transition initiatives

1. THE LIVING COAST
2. THE BRIGHTON ENERGY CO-OP
3. THE BRIGHTON & HOVE FOOD PARTNERSHIP
4. BRIGHTON PEACE AND ENVIRONMENT CENTRE
5. THE WASTE HOUSE

1

©Rick Howorth

The Living Coast: The Brighton & Lewes Downs UNESCO World Biosphere Region

The Living Coast was conceived in 2008 by officers of the Brighton & Hove council. The initiative aims to serve as a world-class demonstration area where humans live in harmony with the local environment. The Living Coast covers almost 400 square kilometres and brings together the three environments of countryside, coast, and city and towns under one united approach. It is led by a partnership of over forty local public, private and civil society actors with a mission to pioneer a positive future that connects people and nature through building relationships and inspiring better ways of living. Activity is focused in projects such as the creation of pilot 'rain gardens' in local parks to help address flood risk and working with local schools to understand the local water cycle. In 2014, the Living Coast secured a Biosphere status award from the United Nations Educational, Scientific and Cultural Organization (UNESCO).

→ https://www.thelivingcoast.org.uk

2

The Brighton Energy Co-op

The Brighton Energy Co-op was set up in 2010 by a small group of local enthusiasts and has now become the largest owner of solar photovoltaic electricity generation in the city (closely followed by the council). Activists were motivated to tackle climate change locally through practical bottom-up action and settled on the local deployment and ownership of solar PV systems as a means of doing so. By the end of 2016 the Co-op had raised more than £1.5 million from local investors to deploy solar panels at eleven sites.

→ http://www.brightonenergy.org.uk

The Brighton & Hove Food Partnership

The Brighton & Hove Food Partnership emerged in 2003 through the joint efforts of local food campaigners, council officers and a national food policy organisation called Food Matters, based in the city. Their aim was to achieve greater co-ordination of activities relating to food in the city. The Partnership is a non-profit organisation and acts as an important information point and conduit between actors in the city. In 2006, a city-wide food strategy – a first in the UK – was developed under the partnership and has since become an important example for other UK cities. As well as working on local policy, the Partnership delivers a range of projects around healthy eating, cooking, growing and composting. More recently the partnership has been involved in supporting local food banks for struggling households and food waste campaigns.

→ http://bhfood.org.uk/

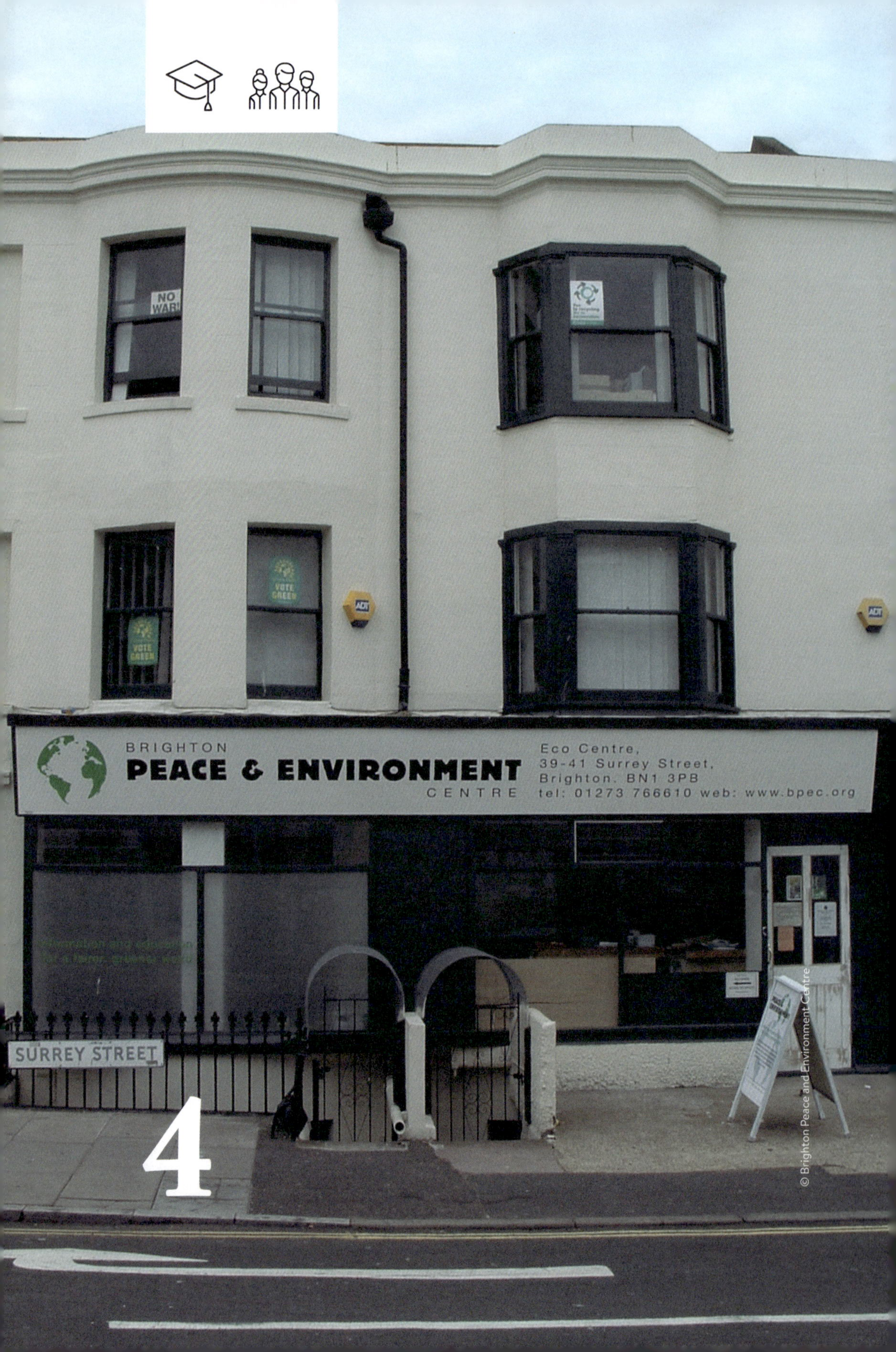
NO WAR!
VOTE GREEN
VOTE GREEN
BRIGHTON
PEACE & ENVIRONMENT
CENTRE
Eco Centre,
39-41 Surrey Street,
Brighton. BN1 3PB
tel: 01273 766610 web: www.bpec.org
SURREY STREET

4

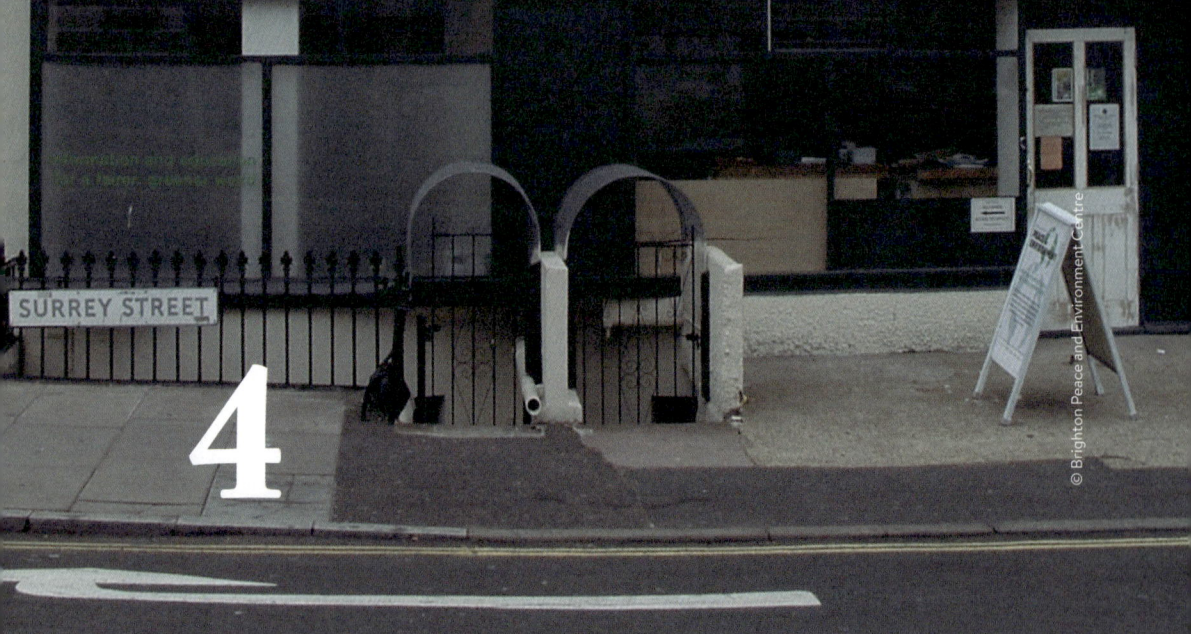

Brighton Peace and Environment Centre

Brighton Peace and Environment Centre began in 1982 with the aim of advancing the education of the public in relation to sustainable development. Their motto is 'Inspiring action through learning about the world'. Their ambition is to help create a just, peaceful and sustainable world and they focus especially on the protection, enhancement and rehabilitation of the environment and on the peaceful resolution of conflicts. The Centre houses resources and provides training and support for school teachers. The Centre also acts as an umbrella body and provides offices and meeting spaces for other local environmental groups and campaigns; it produces the Green Pages (a green business listings service for Brighton & Hove) as a booklet and download; and it runs 'Carbon Conversations', a project which aims to increase public engagement and awareness of the impacts of climate change on people's lives around the world.

→ http://www.bpec.org/index.php

© University of Brighton

The Waste House

The Brighton Waste House was completed in 2014 as a prototype for constructing a contemporary, low energy, permanent building using over 85% 'waste' material drawn from household and construction sites. It is Europe's first permanent public building made almost entirely from material thrown away or not wanted. It was designed by Duncan Baker-Brown, senior lecturer & architect. The construction site was run by national contractors The MearsGroup who also supplied apprentices and Cat Fletcher of FREEGLEUK who sourced material for the build. It was built by over 360 students from The University of Brighton and City College Brighton & Hove. During its construction, it acted as a living laboratory for ecological architectural design and attracted over 750 visiting school children in twelve months. The Waste House is an ongoing live research project, attracting new research grants. However, it is still primarily used as a community hub open for talks and performances while also hosting the University of Brighton's MA in Sustainable Design. The Waste House is also an A* energy-efficient rated building creating 30% more energy than it consumes.

→ http://arts.brighton.ac.uk/business-and-community/the-house-that-kevin-built

± **1,700/km²** urban density

An integrated energy and
climate protection plan

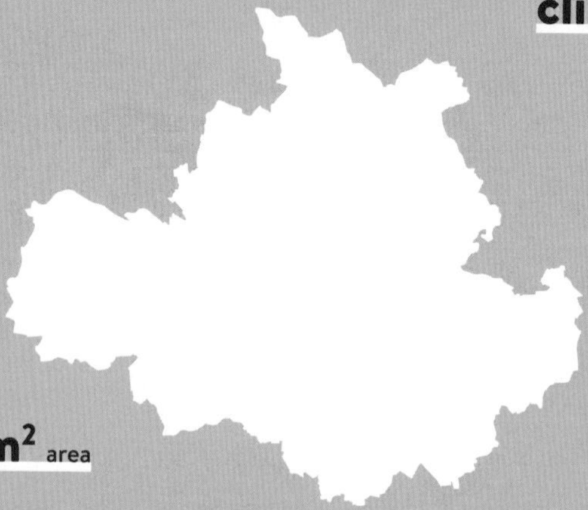

328.8 km² area

± **544,000** inhabitants

Dresden

Dresden is located in Saxony, which is one of the 16 German federal states (Länder). It is situated in Eastern Germany and, therefore, used to be under a communist regime before it gained greater independence in 1990, when Eastern and Western Germany were unified. It has a population of 544,000 citizens and a further rise of 8% is expected by 2030. In socio-economic terms, Dresden is one of the most important locations for business companies in Eastern Germany. These are mainly active in micro-electronics, electro-technology, mechanical engineering or the food industry. The local economy is dominated by small and medium size businesses and a strong service sector.

While the German government is one of the world's frontrunners in addressing climate change, the city of Dresden is trailing behind when it comes to reducing their greenhouse gas (GHG) emissions and increasing their share of renewable energy sources compared to the national averages. However, local initiatives that promote more sustainable ways of thinking, doing and organising are multiplying across the city region. Due to Dresden's communist past and its democratic transition in 1990, a considerable amount of citizens' initiatives thrived in the early 1990s giving rise to meaningful environmental protection initiatives. After this first wave of citizen action, the 2000s witnessed a second wave of local initiatives that now also adopted the idea of sustainable development.

In just over twenty-five years, the city region has thus been exposed to several important transformations in which civil society played a catalytic role. This is still apparent in the many civil society initiatives such as organic food initiatives, local currencies, local transition town and permaculture initiatives and the Umundu festival promoting sustainable consumption for example. However, the local government and business sector have been catching up by establishing and developing initiatives focused on sustainability. For instance, the city is developing a sustainability vision, concrete plans and urban living labs in the frame of the 'Future City Dresden Initiative' (Zukunftsstadt: see www.dresden.de/zukunftsstadt) and works on the implementation of an 'Integrated Energy and Climate Protection Dresden 2030'. Business actors on the other hand are giving rise to new product-service platforms such as sz-bike and the car sharing initiative 'Teil-Auto' for instance, to promote bike and car sharing respectively.

Five inspiring local transition initiatives

1. KLIMASCHUTZSTAB DER STADT DRESDEN
2. COMMUNITY-SUPPORTED AGRICULTURE INITIATIVE SCHELLEHOF
3. VG – USER COMMUNITY FOR ENVIRONMENTALLY FRIENDLY PRODUCTS
4. BÜRGERKRAFTWERKE
5. TRANSITION TOWN DRESDEN

Klimaschutzstab der Stadt Dresden

The city of Dresden is looking towards the future and responding to future energy challenges with its 'Energie fürs Klima – Dresden schaltet' climate protection strategy by implementing energy saving measures, the increase of efficiency and the expansion of renewable energies. To advance this cause, the city established a Climate Protection Unit in 2010 in order to fuel local action to reduce greenhouse gas emissions by at least 10% every five years. The unit focuses on the implementation of the Integrated Energy and Climate Protection Program of the City of Dresden through targeted decarbonisation activities. These encompass the promotion of e-mobility in the city, energy efficiency actions in the building sector, the increase of private photovoltaic modules in the city and the expansion and modernisation of the 550 km district heating system of combined heat and power generation. The unit not only supports the cleantech business scene but also develops campaigns to promote climate-friendly civic behaviour, e.g. by organising the European Mobility Week as a campaign to present sustainable mobility alternatives to local residents.

→ http://www.dresden.de/de/stadtraum/umwelt/umwelt/klima-und-energie/klimaschutz.php

2

Community-supported agriculture initiative Schellehof

Unhappy with the negative environmental impacts of conventional farming, the new generation of farmers of Schellehof set in motion a change process to shift from a degenerative to a regenerative business model. They turned the farm into a biodynamic enterprise that supports the interrelations between soil, plants and livestock. In 2011, they joined the Demeter association, the oldest organic association in Germany and quality leader in this field. In 2014, they decided to establish Schellehof as a community-supported agriculture (CSA) initiative which allows citizens to become shareholders of the farm for a minimum period of twelve months (i.e. a full vegetation period). In turn for their financial contribution, which is payed at the onset of the growing season, participants receive shares of the harvest throughout the year. A CSA business model helps farmers to invest in more sustainable practices up front, regardless of the success of the harvest while it cuts out expenses for intermediaries because it directly links producers to consumers. By 2016, the Schellehof CSA established six food depots (Struppen, Pirna, Leubnitz, Striesen, Plauen and Neustadt) offering organic vegetables, potatos, grain, flour, semolina, bran, a variety of organic animal products (mutton, beef, goose and eggs) and processed foods like bread and roll. Next to life-friendly food production, the initiative also plays an important social role in advancing eco-literacy through reconnecting citizens to nature by organising harvesting, nature conservation and community building activities.

→ http://www.schellehof.de/schellehof.html

VG – user community for environmentally friendly products

The user community for environmentally friendly products VG (Verbrauchergemeinschaft für umweltgerecht erzeugte Produkte) is a large consumer cooperative for regional and organic food in Dresden. With over 10,000 members (2017) the initiative plays a crucial role in offering Dresden residents a relatively easy option to consume sustainable food products and daily goods (detergent, shampoo etc.). Established in 1991 by a handful of engaged citizens, the initiative started off with a small membership base but quickly grew into an association in 1994. As the initiative grew larger, they chose to adopt a cooperative structure in 2005. Food is mainly sourced locally (within a radius of 150 km) to keep the carbon footprint of transport as low as possible and has to follow high ecological standards. To achieve this, the cooperative set up collaborative arrangements with more than one hundred farms and producers in the city region. Most of these collaborations exist for years and safeguard a steady income for organic farmers. In addition, the VG is also well-linked with a lot of other local small producers, helping them to make their sustainable alternatives more known and accessible (e.g. shop for natural colours in Dresden-Striesen). At the moment, the initiative sells eco-friendly food and goods in six different shops, located in different parts of the city. VG members pay lower prices for regional and organic food since there are no more trade intermediaries.

→ http://www.vg-dresden.de

4

Bürgerkraftwerke

Bürgerkraftwerke is a citizen cooperative that brought the idea of civic power plants to Dresden in 2001. This initiative aims to accelerate and scale low carbon innovations in the city by establishing and running renewable energy systems. While their initial interest was to increase civic investments in photovoltaic systems in Dresden, the cooperative evolved towards advancing the field of renewable energy generation in all its aspects: working on the further development of related technologies (e.g. energy storage); overcoming technical and organisational barriers for adapting civic power plants and by fostering the dialogue between all the relevant actors from civil society, city administration and the business sector. In 2016, they ran 79 photovoltaic installations with a total performance of 1313,06 kWp in Dresden and collaborated with, among others, the Local Agenda 21, the foundation German Hygiene Museum Dresden, the DREWAG municipal utilities Dresden Ltd, the SWB Sachsen Solar plc and the environmental centre of Dresden.

→ http://www.buerger-kraftwerke.de

DIE ZEITUNG AUS DER ZUKUNFT

Niederländisch weiter auf DD 2030
denn Vormar...

- 2°C-Schwelle 2018 überschritten
  → Flut seit 2020
- Auswanderung seit 2022
- schon 2015 freie Sprachwahl an DD-Schulen na...
- Griechenlandpleite + Auswanderung
- 2022: Zuwanderungskonzept + Mehrsprachigkeitspro...
  → Zuwanderung wird positiv genutzt
  → neues Hochwasserkonzept fußt auf niederl. Fach...
- 10. Klasse mit niederländisch-Schwerpunkt eröffn...
- kommunale Sprachenkoordinatorin Gundula Schrau...
- Unsere Schüler freuen sich auf die geplante Sprachr...
- ... nach ... - Amsterdam

© Transition Town Dresden

Transition Town Dresden

In 2010, citizens committed to advance sustainability in their city established the Transition Town Dresden initiative (Dresden im Wandel). It is an umbrella initiative that unites a wide range of local projects and initiatives focusing on more sustainable ways of thinking and living. The umbrella initiative is financially supported by two German foundations. Its main aim is to guide the city of Dresden towards a sustainable, post carbon future by raising awareness, showcasing, supporting and developing alternative solutions. Through the umbrella structure, local actors working on sustainability get connected and exchange experiences. The initative also puts the many little but sustainable changes that are taking place in the city region in the spotlight. Transition Town Dresden is part of the global Transition Network, a movement of communities coming together to reimagine and rebuild the world. They focus on a wide range of topics that relate to sustainability in the widest sense, such as permaculture, and cover a variety of low carbon domains, such as water, food, energy, buildings etc. They also pay attention to various aspects of social innovation like reconnecting to yourself, others and the natural world (inner transition), setting up transition enterprises to promote local economic resilience (REconomy) and building a global network through international hubs.

→ http://www.dresden-im-wandel.de

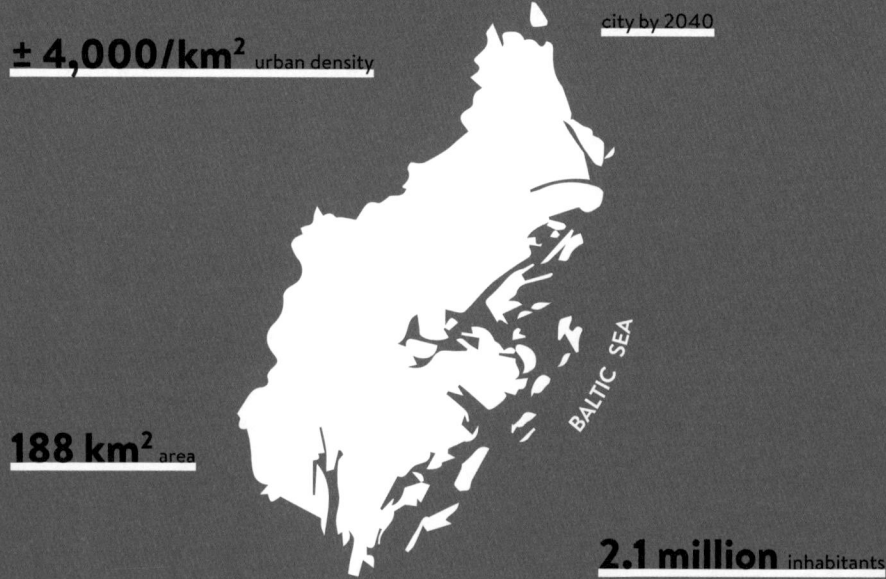

± 4,000/km² urban density

carbon neutral city by 2040

188 km² area

BALTIC SEA

2.1 million inhabitants

Stockholm

Stockholm, the **capital of Sweden**, is situated on **fourteen islands** where **Lake Mälaren** flows into the **Baltic Sea**. It is the most densely populated region of Sweden with more than 2.1 million inhabitants and population size is expected to increase with **30,000-40,000 inhabitants each year**. Stockholm is the **cultural and economic centre of Sweden** and most residents work in the **service, finance or high-tech industry**. The city region comprises **twenty-six municipalities** that have a high degree of autonomy when it comes to decision-making and land use planning. These municipalities differ in area, population size and composition, economy, political set up and priorities, and urbanisation strategy.

Sweden has a long tradition when it comes to sustainability. The nation's sustainability effort has thus evolved over a long period of time and is now well-regulated and formalised. Still, many obstacles are impeding action and only five of sixteen national environmental goals will be achieved by 2020. These do not hamper ambitions, however: Sweden is aiming to neutralise its greenhouse gas emissions by 2045 and the city of Stockholm aims to become carbon neutral by 2040. For many aspects of sustainability – such as climate change, green structure, toxins, biodiversity, waste and water management – there are programs, strategies and plans, both on regional level and on municipal level. More recently, the different governance actors of County Board, County Council and the municipalities are increasing their focus to complex future issues such as water and food supply, climate adaptation, social sustainability and becoming fossil fuel-free.

A wide variety of sustainability initiatives are active within the city region and sprout from public and private sectors and civil society as well as partnerships among those. Many of these initiatives target several challenges simultaneously by combining different aspects of sustainability. For example, the Environmental Workshop Flaten (Miljöverkstan Flaten) initiative unites education, social integration, ecological knowledge and sustainable development by restoring recreation spaces in urban nature. Refo creates jobs involving recycling and thus combines working with recycling, design, education, social inclusion and alleviation of poverty. The Sustainable Hökarängen (Hållbara Hökarängen) covers and combines recycling, cultivation, education, energy and social issues. Likewise, the civil society transition town initiative Värmdö is rolling out a more holistic approach to transition to a low carbon society covering multiple domains such as nature, energy, food and local economy.

Five inspiring local transition initiatives

1. REFO
2. MUNICIPAL ORGANIC FOOD INITIATIVE, SÖDERTÄLJE
3. GREATER STOCKHOLM NATURE GUIDES
4. SUSTAINABLE HÖKARÄNGEN
5. STOCKHOLM GREEN WEDGES COLLABORATION

Refo

Refo, short for Remake and Re-use, is a social enterprize that redesigns and re-uses clothes and items. On the one hand, they distribute secondhand clothes to people in need in the Stockholm region and on the other hand they upcycle apparel and other donated goods and sell these items via a showroom, fifteen small shops and a website. That way, Refo creates social, ecological as well as economical value in an innovative business model. The remaking and redesigning is primarily done by unemployed young people that are outside the conventional labour system. So the initiative generates jobs for those that most need it while their upcycling processes give new life to waste streams in an innovative way bringing design, fashion and waste together in a creative spiral of entrepreneurship. Altogether Refo involves more than one hundred citizens in the business of designing and reselling, crossing the boundaries between for-profit and non-profit.

→ http://refo.nu/

2

Municipal organic food initiative, Södertälje

In 2000, the municipality of Södertälje in Stockholm decided to only serve sustainable and ecologically produced food. The municipal organic food initiative (Ekomat Södertälje kommun) was born and organised directly under the Director of the municipality. A nutritional manager was put in charge of five large catering kitchens. That way, purchases were focused on organic products and an educational approach was enrolled for municipal employees to increase their awareness and knowledge about healthy and sustainable nutrition. Since then, food served at municipal grounds and activities have to comply with six criteria:
- Good and healthy food
- Organically grown
- Less meat, more vegetables and whole grains
- Seasonality
- Locally produced
- Reduced waste

The municipality receives much attention for its work with climate and environmentally friendly food in schools and elderly care homes and since March 2010 all kitchens are eco-certified. The project has spread internationally and has been part of an EU-project called 'Diet for a Green Planet'. Recently, the Södertälje municipality has started a new project 'Matkraft' focusing on sustainable food chains together with the Södertälje Science Park with support from the EU.

→ http://ekomatcentrum.se/sodertalje/

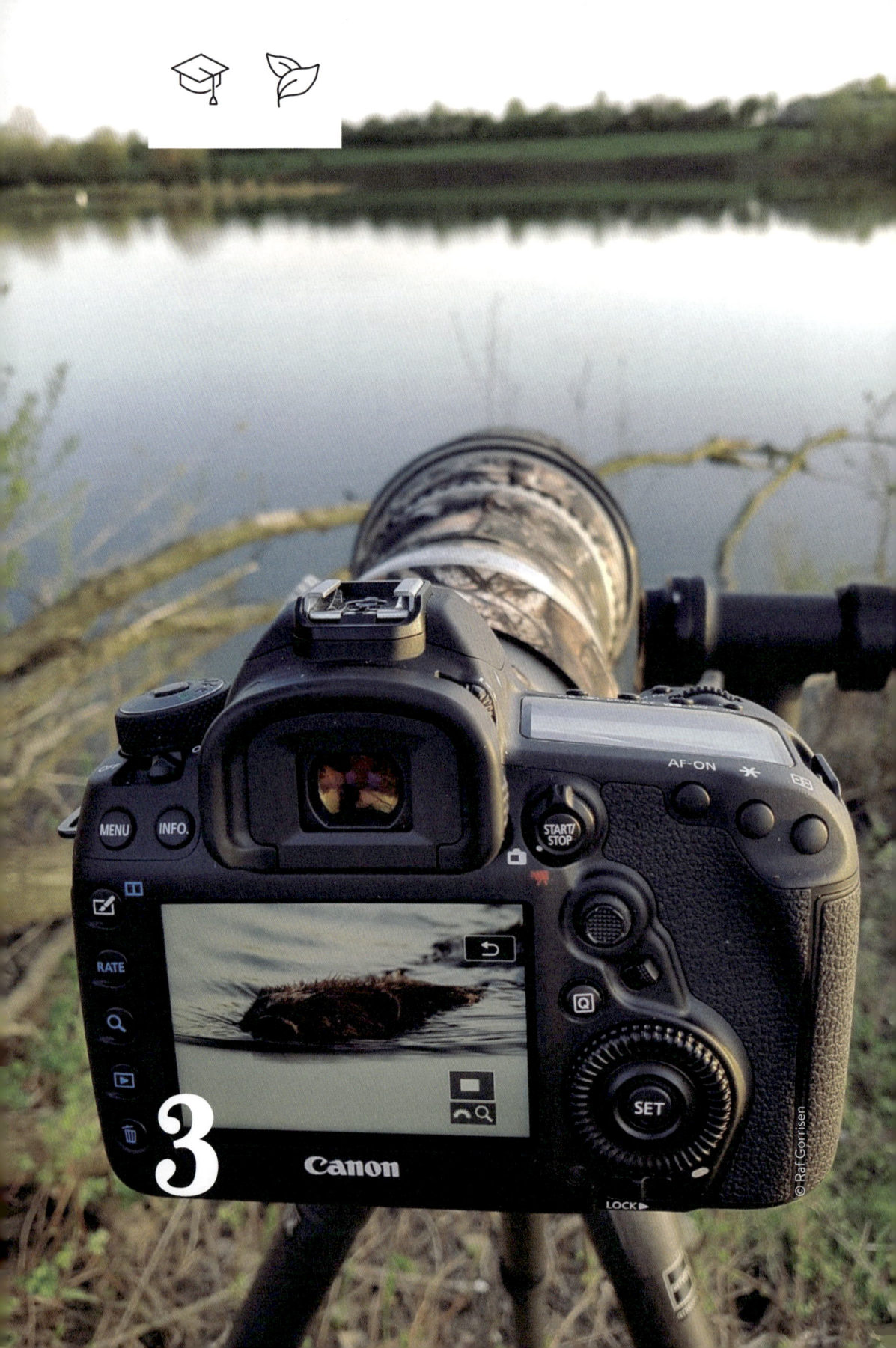

Greater Stockholm Nature Guides

One of the employees of the Stockholm County Administrative Board in charge of developing a visitor's guide and a website for nature experiences within the Stockholm region, realised that there was a need for guidance into nature to make it more accessible to more people. With support from the regional branch of Swedish Association for Nature Conservation (SSNC) in Stockholm, she applied for funding to start a new initiative called 'Greater Stockholm Nature Guides' (Storstockholms naturguider) to educate nature guides. By 2003, a first round of guides was educated and authorized by the Stockholm Visitors Board along with the development of many guided tour manuscripts in various green areas in Stockholm. The initiative aims to reconnect citizens and nature, and its value proposition is simple: if you do not know about nature, it is hard to like and care about it. With this in mind, the initiative aims to raise awareness of the many accessible green areas in Stockholm, as a leverage point to building more general environmental consciousness. The initiative organises year round guided tours for the public. It also organises tours for specific target groups such as unemployed citizens. The guides are authorized as any other guide in Stockholm and get paid.

→ https://stockholms-lan.naturskyddsforeningen.se/om-oss/storstockholms-naturguider/

Sustainable Hökarängen

Property owner Stockholmshem, a company owned by the Stockholm municipality, realised that current practices of property development are no longer appropriate in today's world. In 2011, they launched a pilot initiative in the city district Hökarängen, where they own 3000 apartments and 500 places, to develop the area focusing on sustainability and community building. The company started a citizen dialogue to involve citizens in the decision-making processes. From these dialogues they selected the following core focus areas for action for Hökarängen: energy efficiency, centre area development and artistic businesses, behavioural change and climate change adaptation. The initiative brought together a variety of innovative actors to explore how to engage local citizens in sustainability issues. As part of the district lies in a deprived area, the initiative also sought to promote social inclusion and spread sustainable alternatives within the municipality at large. The Sustainable Hökarängen initiative sparked citizen engagement by bringing them together around long-term envisioning and more holistic planning approaches. It even sparked a new initiative, the Hökarängen transition and permaculture project (Hökarängen Omställning och Permakulturprojekt; HOPP!). The HOPP! initiative received start-up funding from Sustainable Hökarängen and organised transition dialogues with citizens, courses for residents around community sustainability practices and permaculture gardening. The HOPP! initiative however was not able to acquire enough funding to sustain their activities beyond 2016 and has since been dormant.

→ http://hallbarahokarangen.com/

Stockholm Green Wedges Collaboration

In 1996 a collective of NGOs, representing environmental protection, tourism, outdoor recreation and cultural heritage, established a new network to better safeguard the natural landscapes surrounding Stockholm. Fuelled by the concern that increasing urban development should not go at the expense of vital green infrastructure in the city, the collective gave rise to the Stockholm Green Wedges Collaboration (Samverkan Stockholms gröna kilar) initiative that sets out to protect the green wedges of Metropolitan Stockholm from urban development. This collaboration aims to bring together all municipalities bordering on the green wedges in a unified approach to coordinate strategies and action for the protection of the green infrastructure. As a first step in 2006, the six municipalities and NGOs related to the Rösjö green wedge were invited to a workshop by the regional branch of Swedish Society for Nature Conservation in order to form a mutual understanding of the green wedge as an ecological unit and to inspire cross-border/sector as well as municipal/civic society collaborations. Following that workshop the first activities included joint study visits, the creation of a visitor's map, and building networks with regional authorities and other external key actors. Eventually a more formal platform for collaboration was created that also gained political support – the Rösjö Green Wedge Collaboration (Rösjökilen samverkan) initiative. This more formal initiative raised attention in the other green areas of the Stockholm region and inspired similar action in another green wedge. What is more, both the civic and public initiative paved the way for the creation of a regional green wedge council, which is coming closer to realization.

→ https://stockholms-lan.naturskyddsforeningen.se/natur-grona-kilar/

± 3,351/km² urban density

A civic movement to create a **resilient, liveable** and **bike friendly** city

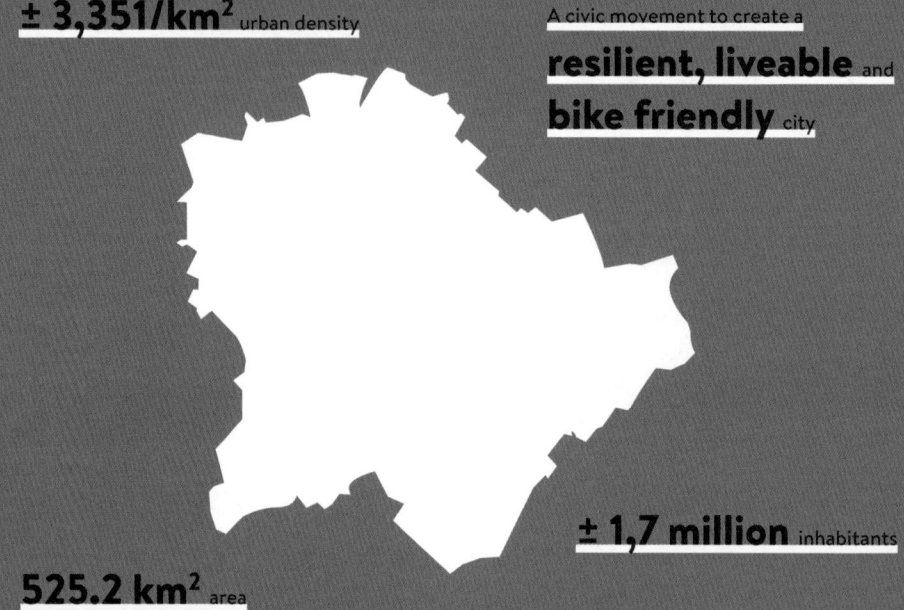

± 1,7 million inhabitants

525.2 km² area

Budapest

Budapest is the **capital** of Hungary, situated on both sides of the **river Danube**. It is the most important **social, historical, political, trade, traffic centre** of the country. It houses approximately **1.7 million** inhabitants (2016) and longer term forecasts predict a decline in population growth in the future. Hungary is a post-socialist state and a more recent member of the European Union (EU). While environmental issues in Hungary have generally grown in importance since the regime change, this has been within a generally unpredictable governance context. Since **EU accession in 2004**, several policy and regulatory changes have been achieved in order to comply with **EU environmental legislations**, linked to funding and strategic development programs. However, environmental and sustainability issues have never been sufficiently mainstreamed, and restructuring processes within national governance have further limited sustainability approaches in public policy in favour of economic growth.

Because of this, civil society organisations promoting sustainability still have to deal with an unstable operating environment which prevents them to climb out of the margins of current governance and thus limit their impact. In response to this, a diverse civic-initiated resiliency movement has developed recently. This movement has increased environmental consciousness locally, and has inspired new visions and implementation strategies for a more socially active and liveable Budapest, supported by residents. From urban mobility to local agri-food systems, and increased citizen awareness of environmental issues, both the infrastructure and the consciousness of the city and its inhabitants are in flux. Positive examples of both established and emerging transition initiatives abound, dealing directly with the unique conditions of Budapest and providing insights in how a bottom-up movement is promoting more social and environmentally friendly ways of thinking, doing and organising.

Five inspiring local transition initiatives

1	SZATYOR ASSOCIATION
2	TRANSITION WEKERLE
3	CARGONOMIA
4	HAZIKO
5	MAGNET COMMUNITY BANK

Szatyor Association

Szatyorbolt (Szatyor Egyesület) is a new social enterprize that offers shopping alternatives for citizens who care about their environment, health and community. It is a local community food cooperative in Budapest, that promotes eco-consciousness, community building and sustainable living. They do this by supporting short chains between producers and consumers, by supporting collective learning about life-friendly food in their community, by sharing information with other food communities in Budapest and by networking activities to build eco-literacy. For instance, they collaborate with the Hungarian Association of Conscious Consumers, to spread knowledge about community driven food systems and they often participate in outreach events organised by university research groups and local sustainability oriented NGOs to provide the perspective of a community driven conscious food initiative.

→ http://szatyorbolt.hu

2

Transition Wekerle

Transition Wekerle (Átalakuló Wekerle), housed in Europe's largest Garden City, was initiated by a few engaged citizens that wanted to rejuvenate the area both socially and ecologically. In 2012, they became the first official transition town group in Budapest with a regular representation within the international transition network movement. They focus on strengthening community bonds while promoting the reconnection of citizens to nature for which Wekerle is ideally situated being such a green Garden City. Their successes have influenced citizen groups working to increase social cohesion in neighbouring Budapest districts. Members are often called upon to share organisational and campaign strategy tips with similar transition themed groups in other Budapest districts. Their model has shown that non-hierarchical, horizontally organised groups with fluctuating members and causes, which remain linked by a number of dedicated vision makers, can remain effective, relevant and impactful despite changing dynamics. Their pioneering approach to create an open dialogue with the municipal government inspires many other community initiatives to participate in local governance to help rebuild trust between civic and public instances.

→ http://atalakulowekerle.blogspot.hu/

Cargonomia

Cargonomia is the formalisation of a pre-existing cooperation between three socially and environmentally conscious small enterprises operating in or near Budapest. Partners within the project include the Do it Yourself Bicycle Social Cooperative (Cyclonomia), an organic vegetable farm and sustainable agriculture community education center (Zsamboki Biokert) which distributes weekly vegetable boxes to food communities in Budapest and a self-organised bike messenger and delivery company (Kantaa). Cargonomia is the node between the activities of the partners that brings together sustainable food production, promotion of low carbon transport solutions and bicycle competency advocacy in a mutually beneficial way. Next to offering low carbon and environmentally friendly food and mobility services to citizens, they also strengthen the bonds between responsible urban and rural initiatives and actively engage in the transition processes in the city of Budapest. They organise community forums, support the start up of other initiatives and engage in dialogues with public authorities to advance progress on sustainability in their city. Team members were some of the main organisers of the fifth International Degrowth conference, during which a special session was held for municipal representatives and decision makers titled 'Degrowth in Parliaments'.

→ www.cargonomia.hu

Haziko

Haziko is a catering company which sources only local and mostly organic ingredients directly from producers in and around Budapest. Haziko aims to serve as the missing link between rural producers and urban consumers. By indicating the sources of all ingredients on biodegradable packaging of products, they aim to ensure full transparency of the complete food chain. The catering company aims to serve as a pioneer in the budding conscious food scene in Budapest, while also having a rural development consultancy branch, Agri-Kulti, which regularly takes part in academic research projects and conferences related to sustainable rural development. They have been invited to numerous forums organised by the Ministry of Rural Development to share their ideas on more supportive connections between urban consumer bases and rural producers. They have developed strategic partnerships with community-oriented businesses such as Magnet Bank, and have recently established a number of cafés in the city offering locally sourced food.

→ www.haziko.farm

Magnet Community Bank

The Magnet Community Bank was established more than twenty years ago and has since won many awards for being socially responsible. In 2010, the bank reorganised its operations to increase transparency, sustainability and social investment in local projects. Magnet Bank's headquarters serves as a community centre which regularly hosts discussions and workshops related to community building and sustainability initiatives. The bank has focused its efforts to provide banking support directly to civil initiatives. They face great pressure within the overall banking sector, but try to initiate a more community centered shift in banking and investment through their own example. An example of this is the community giving program, which allows customers to transfer a portion of interest earned on their individual accounts to civil initiatives in Hungary.

→ www.magnetbank.hu/en

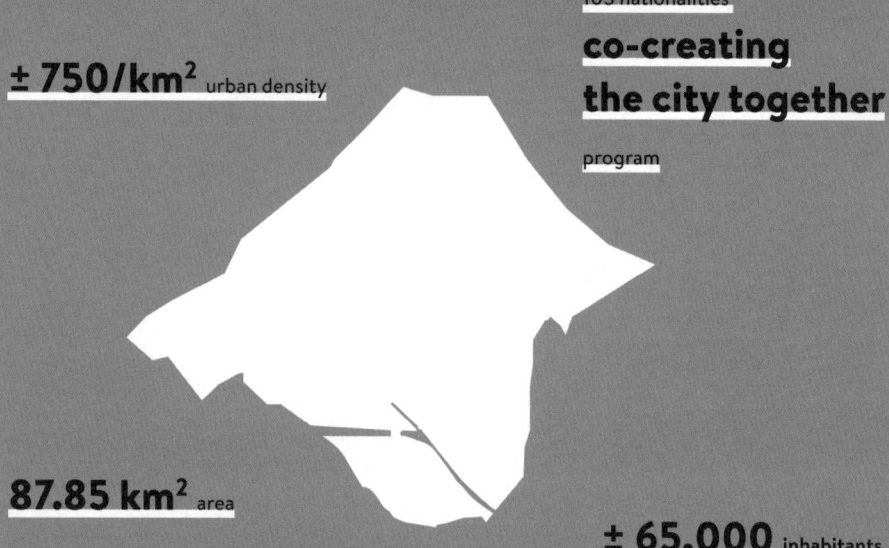

± 750/km² urban density

103 nationalities
co-creating the city together
program

87.85 km² area

± 65,000 inhabitants

Genk

The city of Genk is a **medium-sized** Belgian city, located in the Flemish speaking Eastern part of Belgium and houses **65,000 inhabitants**. Over the past one hundred years, Genk's history has been shaped by **coal mining**, which led to a rapid period of **migration and industrialisation**. This shaped the socio-economic fabric of Genk, attracting more **industries**, giving it a strong **working class** background and a **diverse population**. Genk thus shares the characteristics of European post-industrial cities that have gone through economic restructuring and are searching for a **new identity**. Economically, Genk is the third most significant city in Flanders with most jobs in the **chemical, heavy and service industry**. In 2014, the largest employer of the city, Ford Genk, closed down, giving a severe blow to the local economy.

Due to its specific history and economic restructuring, Genk is rethinking its economic backbone and actively exploring new economic notions such as cleantech, circular economy and smart city, to regenerate a more resilient local economy. Examples include developing Genk as a hub for the circular economy (RE-MINE Poort Genk Project) and stimulating industrial eco clusters and reducing environmental impacts via the Collective Cleantech Realisation Plan. The city has been at the forefront of experimenting with social innovation and inclusive policies. For instance, current policy programs have been inspired by a large civilian brainstorm and the city puts a lot of focus on participation and empowerment.

Today, no less than 103 different nationalities live together in Genk and it is the only Flemish city where people with foreign roots are in the majority: 54% of the residents have immigration backgrounds. Consequently, the city houses an energetic and varied local social scene that is underpinned by a huge diversity of associations and clubs, organising a wide range of activities. The city also has a dynamic volunteer base active in environmental sustainability: a multitude of organised volunteers (e.g. Natuurpunt Genk, Velt Genk, Heempark, Mijnboretum Bret, Tuin van Betty) play a significant role in the management and conservation of Genk's natural capital. Both district managers and community builders play an important role in connecting change makers and developing and supporting initiatives. What is more, given its focus on social innovation, the city and its citizens are experimenting with new governance models such as public-civic partnerships to advance mutual benefit creation via collaboration.

Five inspiring local transition initiatives

1	BEE PLAN
2	ECO COMMUNAL GARDENS
3	PASS-ON SHOP
4	RE-USE SHOP
5	HEEMPARK

Bee Plan

In May 2013, two engaged citizens displayed the documentary *More than Honey* at a local open environmental council meeting in Genk upon which the first Alderman facilitated a brainstorm with the sixty people present on how they could transform Genk into a bee-friendly city. A multi-actor working group encompassing citizens, beekeepers, city services and environmental organisations, was established in a next step to develop an urban 'Bee Plan' for the city. Nine months later, the Bee Plan was approved by the bench of aldermen and implementation soon followed. By changing public green management approaches such as completely banning toxic pesticides, planting borders with nutritious flowering plants and increasing appropriate housing for bees (bee hotels and beehives), the city administration changed its practices towards more bee-friendly ones. Awareness rising activities include organising events such as bees in the library, at school and by putting honey in the picture at local restaurants. A dedicated team of voluntary bee ambassadors (the Bee Team) visit local events, handing out bee-friendly flowers or seeds to citizens, thereby spreading bee-friendly thinking and doing to the wider public. The city allocated extra resources to promote urban beekeeping, which germinated many more initiatives such as the Bee Collective in Zwartberg that demonstrates the importance of bees as pollinators via their 'bee tower' public demonstration exemplar and trains local beekeepers.

→ http://www.heempark.be/Genks_bijenplan

Eco communal gardens

Allotment gardens have long been present in the mining districts of Genk. The past decades however, increasing abandonment led to the decay of these communal areas. In order to regenerate the community, a couple of engaged citizens started a renewal project in the district of Sledderlo to turn these abandoned plots into vibrant, social and organic food gardens. They soon acquired full support of the city to regenerate these areas. Inspired by community building initiatives they called it 'Together Gardens' (Samentuinen). Thanks to this public-civic collaboration, involving the local intermunicipal waste company (Limburg.net) and the non-profit guild for eco-friendly living and gardening (Velt), the initiative managed to get funding from the Flemish government. In 2005, the first twenty-four plots were transformed into organic vegetable gardens. Velt organised training courses around eco-friendly gardening and Limburg.net promoted circular economy thinking. This initiative inspired another group of citizens, located in Genk-Noord, to do the same in their district and the city administration helped them to secure the necessary funds. In 2011, the successful multi-actor alliance transformed another plot of decayed allotment gardens into organic gardens. The renewal approach was so successful that the Flemish government decided to set up a fund to replicate this eco-friendly collaborative model of 'Together Gardens' across the whole of Flanders. In 2017, almost one hundred of these gardens have been developed in Flanders.

→ http://www.samentuinen.org/

Pass-on shop

A local resident of the city district of Waterschei in Genk established the Pass-on shop initiative (Het halve eurootje) in 2009 to promote sustainability and social cohesion in her community. The Pass-on shop collects used apparel and resells these for half a euro. It thus not only supports re-use, it also offers apparel to low-income households at low prices. The shop is also a meeting point for local residents and regularly organises workshops about sewing and cooking to bring together a diversity of locals in a positive atmosphere.

The start-up of the shop was financed via the district budget of the city that aims to seed bottom-up initiatives and one of the community builders of Rimo gave administrative support. Currently, the shop is run by seven volunteers, four of them since the start-up in 2009. Part of the revenues are invested in other socio-environmental charities. The Pass-on shop model has been replicated to Sledderlo and Winterslag, two other districts in Genk, with varied success.

Re-use shop

The re-use shop 'Reset' in Genk is a social enterprise that promotes re-use of goods. In Flanders, product re-use is mainly organised by the re-use centres, non-profit social enterprises that collect and repair donated goods such as furniture, equipment, toys, clothes etc. for resale as secondhand goods in their chain of re-use shops (kringwinkels). By offering employment for unskilled personnel and providing quality goods at low prices for low-income households, the re-use centres play an important role in the local social economy. The re-use initiative was established in 2002 and has since then established more than one hundred re-use shops in Flanders, allied with thirty-one re-use centres. In 2015, this Flemish non-profit organisation collected over 69,550 tons of used products and recorded 5.6 million transactions with customers. They thus significantly reduce waste (and carbon emissions) while supporting the social economy. Over 80% of the more than 5353 people working in the organisation are long-time unemployed or have a low education level. They are co-funded by the Flemish government, at first for their support in the implementation of waste policy, nowadays because they support product and materials re-use whilst supporting the social economy in Flanders.

→ https://www.dekringwinkel.be/centrum/7/winkel/47/index.html

Heempark

The Heempark, nowadays a public-civic partnership, combines city-supported and volunteer activities focused on eco-friendly land management practices. It originated in 1991 as an initiative of a group of local volunteers aiming to safeguard the local agricultural heritage through the Heempark model – a combination of demo sites of local agricultural practice with a clear focus on environmental sustainability. As the initiative grew more popular and attracted more visitors than the volunteers could handle, the city stepped in to offer support in terms of personnel and infrastructure. Via the establishment of the urban Environment and Nature Centre at the Heempark, that operates under the Department of Environment and Sustainable Development of the City, the city helps with the coordination of the educational and demonstrative activities. The volunteers maintain the different demonstration gardens and organise events to promote eco-friendly practices, while the city facilitates approximately 350 educational groups. The Heempark initiative is an important node in the city that brings together most initiatives involved in environmentally friendly land use management and is an important hotspot for citizens, attracting more than 10,000 visitors yearly. Working together, the civil servants and volunteers organise a wide diversity of activities, including sessions about herbs, bees and vegetables, organic food cooking classes with autistic children, energy cafés and keynotes around innovative themes such as biomimicry.

→ http://www.heempark.be/

PART TWO:

HOW CHANGE MAKERS ARE ACCELERATING URBAN SUSTAINABILITY

Chapter 4:
Mechanisms for acceleration

Never before have the possibilities for action been so abundant. Never before has the potential to interconnect all these actions been so great. Therefore the time for putting the blame to those in power lays behind us and the time for kick-starting small but massive action lays in front.

Thomas Lommée, Designer

MECHANISMS FOR ACCELERATION

Replicating

Replicating is the take-up of new ways of thinking, doing and/or organising of one transition initiative by another transition initiative or different actors in order to spread these alternative ways. In this, urban change makers play a vital role in catalysing the genesis of alternative initiatives in their cities. They reproduce life and eco-friendly initiatives that originated elsewhere and often tailor these to their local conditions. That way sustainable ways are not only transferred from one region to the next, they are often place-based interpretations that involve widening circles of people over time.

How this mechanism contributes to urban sustainability

Through replication, initiatives draw inspiration and transfer concepts and ideas on sustainable alternatives from initiatives outside their location. Formats can be copy-pasted from one place to another, reproducing identical initiatives. Generally though, they are tailored to the local context to be more aligned with the place-specific characteristics or conditions. Via this mechanism, positive change is scaling geographically as more and more citizens become involved in more and more initiatives worldwide. Replicating can allow quicker uptake and spread of more sustainable products, practices and services because it can reduce learning curves when experience is shared.

Findings

Replication is a widespread phenomenon and generally adds to an increase in the diversity of sustainable alternatives in cities. From our experience and interaction with initiatives, we learnt that there are many enablers to replication: social media, open source platforms, informal networks between initiatives and festivals or conferences open to initiatives where ideas are openly shared and celebrated. Change makers are pivotal in the transfer of ideas, practices and tools. Just like bees, they pollinate their local environment with novel ideas that can flourish when the right conditions are created. They are thus important catalysts of open innovation in cities.

Example

Striking examples to illustrate how more sustainable ways of thinking, doing and organising are multiplying across the globe are the transition town initiatives **Dresden im Wandel**, **Transition Värmdö (Omställning Värmdö)** in Stockholm and **Transition Wekerle** in Budapest[9] that were all inspired by the transition town of Totnes in the UK. In 2006, two citizens of Totnes kickstarted a community renewal project on how they as a community can address the challenges of peak oil and climate change. The community self-organised into groups concentrating on food, transport, energy, business and livelihoods, health and wellbeing, building and housing, and inner transition. In a co-creative setting they explored how they can change the world street by street and how they can build local resilience by rethinking their economy. This led to many projects such as Grown in Totnes, Keeping Totnes Warm and REconomy, in which they explore relocalisation as a community-led developmental strategy to transition to a more resilient and sustainable community. The novel ways of organising and facilitating change quickly spread across the country and across the globe, seeding similar initiatives in Dresden, Budapest and Stockholm. In 2016, there were 1200 transition town initiatives established in 48 different countries[10]. This example also illustrates the importance of place when it comes to replication. The transition town model is not a solution to be copy-pasted elsewhere. Rather, it is a set of values, principles and processes that can contribute to building community resilience in a way that is place-based.

MECHANISMS FOR ACCELERATION

Partnering

More sustainable practices and ways of thinking diffuse better through well-connected urban social systems. Urban change makers can stimulate connectivity by establishing partnerships between initiatives or between initiatives and other urban actors. Through partnering, initiatives can create interrelations with important key partners and thereby open up new opportunities for establishing synergies or new channels for diffusion of alternative ways of thinking, doing and organising.

How this mechanism contributes to urban sustainability

Partnering occurs in a great variety of ways. Partnerships can be short-term, long-term, informal or formal, ad hoc or strategic. Initiatives can join up to co-organise activities and events, to share knowledge and expertise as well as complement each other in skills, know-how, resources or infrastructure. Via partnering, initiatives aim to unlock new ways of reciprocal value creation by promoting synergies. Or simply put, partnered initiatives can create more positive impact together than they would separately. For instance, through partnering, initiatives can improve their strategic agency, legitimize new governance models such as public-civic partnerships or support higher levels of coordination and aggregation. Consequently, partnering endorses both the speed and scale of positive change since it promotes the self-organisation of open innovation in cities. What is more, unusual partnerships are often catalysts exploring and developing new logics of value creation and establishing new and more sustainable value chains in cities.

Findings

In the five city regions under study, partnering occurred in a myriad of ways. In its most simple form, partnering occurs when two initiatives join forces. This way they can bring together different communities or share resources or infrastructure. When more initiatives partner, they can form strategic alliances that enable them to coordinate activities on a higher level and increase their visibility, legitimacy and influence. Ultimately, partnering lays the groundwork for new innovation networks to arise as it enables the interconnection of a multitude of partners across domains and sectors.

Example

A striking example of partnering is the **Heempark** in Genk. Originally established as a civic initiative to safeguard sustainable agricultural heritage, it rather rapidly evolved into a public-civic partnership combining city-supported and volunteer activities. Both parties collaborate in a mutually beneficial way to promote and scale sustainable ways of land management and food production. In this partnership, volunteers invest their time and expertise in managing the gardens in an eco-friendly way and in involving the community while the government invests in infrastructure and coordinating personnel. Working together like this allows them to involve much more citizens in their activities and attract more visitors to their premises. Without this formal partnership, the Heempark would not be able to accommodate 350 educational groups and manage 10,000 visitors a year. Thanks to this synergism, the intiative also became a hotspot for partnering, bringing together initiatives active in nature conservation, sustainable living and sustainable food production. Through partnering, transition initiatives can seed wider processes of change by involving public bodies in more sustainable ways of thinking, doing and organising. Public bodies can then help mobilize resources and develop governance settings that are favourable for the further promotion of sustainability.

MECHANISMS FOR ACCELERATION

Upscaling

Accelerating urban sustainability requires the mainstreaming of sustainable ways of thinking, doing and organising in the city. For that to occur, new ways of thinking, doing and organising need to be scaled up. Local transition initiatives can promote upscaling by increasing the amount of people involved in sustainable alternatives.

How this mechanism contributes to urban sustainability

This mechanism contributes to acceleration of sustainability because more people involved in alternatives means that the number of sustainable products, practices or services offered in the city increases. The growth in number of users advances exposure and the uptake of alternatives because it adds to increased credibility and legitimacy. And more members allow the initiatives to increase their activities such as demonstration sites or events showcasing sustainability exemplars. This again helps to attract more citizens to these activities. Growing beyond a certain threshold helps to become more influential to impact political agenda setting. What is more, upscaling can also prompt positive spillover effects, transforming the wider system when sustainability aspects spread to other areas of the value chain.

Findings

Upscaling is a commonly observed phenomenon across the city regions and is most prominent in the food domain. However, transition initiatives are often critical towards upscaling because they experience a tension between the management of quantity and the management of quality. Some initiatives explicitly do not want to grow in numbers but rather in the quality of the services they provide, so they opt for replication rather than upscaling, often supporting other change makers in establishing similar initiatives in other parts of the city. This highlights that not only the size but also the direction (quality) of growth matters for upscaling. In other words, transition initiatives put much more emphasis on quality than quantity, shifting value regimes from 'more' to 'better'. Our findings also indicate that political backing – providing time, people and resources – is not a necessary condition for the upscaling of sustainability. Transition initiatives have extended their memberships even without such endorsement. But while upscaling of sustainable approaches is happening in each of the five cities, the research also reveals 'limits to growth', meaning that certain initiatives appear not to be able to grow beyond a certain number of citizens involved. Here, political support could contribute substantially to upscaling.

Example

One of the most striking examples of upscaling is the **VG consumer cooperative** for regional and organic food in Dresden. It started with just a handful of active members in 1991 and is nowadays one of the biggest consumer cooperatives throughout Germany with more than 10,000 members in January 2018. But it becomes really interesting when we look at the dynamics of change that this sustainable food cooperative initiated: it created a ripple effect of positive change across its entire supply chain. Because of the constant high demand for eco-friendly goods, about one hundred farmers in the city of Dresden switched from conventional to organic farming. The long-term cooperation arrangements with VG helped farmers and food processors to make the necessary investments while guaranteeing a steady income. This reciprocal relationship between the food cooperative and the farmers created favourable conditions for the farmers to upgrade their production to more life-friendly practices. A process that is much more difficult to go through at normal market conditions.

MECHANISMS FOR ACCELERATION

Instrumentalising

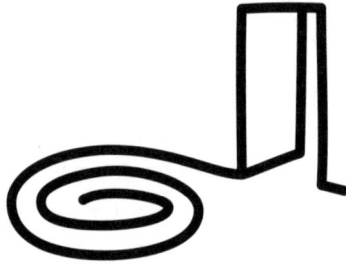

Transition initiatives can increase the pace of positive change in their cities by seizing relevant windows of opportunity at local, regional, national or international governance levels. This means they can tap into and capitalize on external trends and (funding) opportunities to support new ways of thinking, doing and organising locally. Or, put differently, resourceful change makers know how to navigate a landscape of opportunities and sail the waves of change to strengthen the sustainability work of their initiatives locally.

How this mechanism contributes to urban sustainability
By capitalising trends and opportunities, local transition initiatives align local change processes with regional, national or global change dynamics and mobilize external support to fuel local action. These can encompass financial resources but also access to knowledge, skills and networks.

Findings
Across the city regions, findings show that many transition initiatives have successfully seized windows of opportunity and mobilised resources to advance sustainability locally. In doing so, they have drawn on multiple sources of external support. These have been resources such as finances, but also the provision of advice and guidance, or the provision of institutional and physical 'free spaces' for experimenting with sustainability solutions. Some initiatives also used their strategic agency to advocate and lobby for political leeway to innovate. In other words, transition initiatives can also create opportunities that allow sustainable alternatives to enter the political narrating space and agendas. Most often however, initiatives leverage opportunities provided by the local, national or European level to acquire the necessary resources to maintain and scale positive impact locally. Several of the transition initiatives under study have profited from progressed urban planning and decision-making, especially if these have been underpinned by genuine sustainability ambitions and long-term visions.

Example
The Brighton & Hove Food Partnership, set up in 2003 by a number of individuals working within the voluntary sector that felt there was a need to achieve greater co-ordination of activities relating to food, has a good track record of seeing national level policy developments as opportunities which they can harness to strengthen progress towards sustainable food systems within their city region. For instance, when the government introduced the School Food Plan in 2013, the Partnership viewed this policy plan as an institutional space for them to work within as they pursue ever better environmental standards for food served in schools. Thanks to a clear strategy and a forward planning approach, the initiative was able to respond proactively to developments to sustain and further their mission, for instance by tapping into national level funding agendas to promote health care, food security and reduce childhood obesity. The initiative does not directly seek to influence national policy but instead engages with national-level bodies, such as Sustain, in order to feed into and support their work – e.g. the current Campaign for Better Hospital Food. As such, they tap into national change dynamics to advance their work locally which enabled them to create the first ever UK city food strategy, that is now being replicated to other cities in the UK.

MECHANISMS FOR ACCELERATION

Embedding

Embedding means that new ways of thinking, doing and organising get integrated in city-regional governance patterns. Or put differently, that more sustainable alternatives get ingrained in the routines, habits and customs of citizens and in the political and institutional structures that encompass rules, legislation, policy and processes.

How this mechanism contributes to urban sustainability

Acceleration of urban sustainability occurs when sustainability aspirations are genuinely anchored in the vocation of the city. This requires both institutional innovation, on the local and other levels of governance, and the widespread adoption of sustainable values and practices by the general public. Transition initiatives support the mainstreaming of new ways of thinking, doing and organising by showcasing that sustainability is attainable and by involving the wider public in urban renewal projects. Local governments on the other hand, can integrate more sustainable ways of thinking, doing and organising into their city making processes, setting the example and promoting legitimacy of sustainable alternatives. They can also facilitate participatory innovation spaces to support the continuity of urban social innovation by empowering collective learning and action. Most importantly however, local governments hold the key to redesign urban institutions to promote more integrative and transformative thinking so that more integral solutions and faster uptake of alternatives is enabled. However, most institutions are still very much compartmentalized structures that reinforce reductive and silo thinking. They have been designed by a logic that did not factor in sustainability, which means they locked-in unsustainability and are therefore not well-equipped to incorporate novelties. Transition initiatives challenge this outdated logic. By establishing collaborative arrangements of co-evolution, local governments and transition initiatives can accelerate the pace of positive and meaningful change in their cities.

Findings

Combining the insights gained across the five city regions demonstrate that embedding occurs in multiple forms: (1) embedding new ways of doing (practice), (2) embedding new ways of thinking (culture) and (3) embedding new ways of organising (structure). Our findings also show that both local governments and transition initiatives are instrumental for mainstreaming more sustainable alternatives. Embedding occurs via routinisation, when sustainability becomes the norm and via institutionalisation, when sustainable alternatives are integrated into urban establishments. Both collaboration or confrontation appear to promote embedding. In the cases where transition initiatives offered new solutions to problems of local governments, state actors aligned their activities with them based on collaboration. In such contexts, initiatives have often had to follow a strategy of 'fit & conform'. By contrast, where transition initiatives challenged the stakes and interests of incumbent actors, embedding generally involved a process of confrontation and contestation. Here, transition initiatives' strategies were much more of a 'stretch and transform' nature. However, embedding proved to be easier if both transition initiatives and incumbents perceived a synergy between the objectives and the activities they pursued.

Example

Initiated by a collective of NGOs, the **Stockholm Green Wedges Collaboration** (Samverkan Stockholms gröna kilar) initiative brought together all municipalities and NGOs of the Rösjö green wedge. The aim was to develop a mutual understanding of its role as an ecological unit and to inspire cross-border/sector as well as municipal/civic society collaborations to prevent urban development from eating up this important green refuge in the area. The process gave rise to a more formal platform for collaboration that also gained political support, called the Rösjö Green Wedge Collaboration (Rösjökilen samverkan) initiative. Thanks to this support, the green wedge collaboration was gradually incorporated in the official work profile of civil servants and budgets were allocated around shared objectives. So the initiative succeeded in integrating a more holistic approach to address nature as an important part of human urban development into official programs. The initiative also paved the way for a regional green wedge council, which is coming closer to realization. This example illustrates how collaborative arrangements can support embedding processes. In contrast, the **Hungarian Cyclists' Association** had to develop a more confrontational strategy to promote embedding of this low carbon transport mode in the city of Budapest. As city policymaking remained dominated by the logic of putting the cars central, they had to intervene in the city in a way that caught the attention of the politicians, and reclaim space for bikes. They achieved this by organising Critical Mass bike events that mobilized tens of thousands of bicycle enthusiasts to take over the streets. Their creative and elegant solution provided them dialogue space to participate in policy debates and convinced the ruling government to invest in making the city more bike-friendly.

Chapter 5:

Overview of sustainability dynamics in five European cities

Sustainability of the planet is not merely about being a good citizen and recycling; it is ultimately about maintaining an intimate relationship with nature. Research shows that in order to truly care about 'being green', one must actually have meaningful exposure to nature.

Eva Selhub, Physician and Alan Logan, Biophilosopher

How transition initiatives are replicating sustainable alternatives

We observed different dimensions of replication: inward, outward and internal replication. Inward replication is when sustainable alternatives from elsewhere are transferred to the city region. For instance, the initiatives Dresden im Wandel, Transition Wekerle in Budapest and Transition Värmdö (Omställning Värmdö) in Stockholm were inspired by the transition town movement that originated in Totnes in the UK. Once established, a new transition initiative can further fuel outward replication spreading sustainable alternatives to other cities. For instance, one of the activities organised by Stockholm's Värmdö transition town initiative called 'A week for the future' (Framtidsveckan) has been replicated to fifteen different locations in Sweden after it received national interest. The activity brings together urban actors to explore and share a wide variety of sustainability themes while fuelling networking activities for local action. So replication of one initiative in a city can induce a cascade of change in the wider system it is nested in. Another illustration of replication is the CSA farm Dein Hof in Dresden which was founded after an intense exchange process with members of the German-wide network for community-supported agriculture.

Another example how one initiative inspired replication to other cities is the Eco Communal Gardens initiative (Samentuinen) that has been pioneered in Genk in 2006. The initiative promotes communal gardening in an eco-friendly way. Their socio-environmental approach developed in Genk was successful in simultaneously regenerating degraded plots of land and promoting social cohesion of its multicultural communities. Impressed by this approach of co-creation between civil society and public institutions, the Flemish government set up a financial support system to promote the further replication of this initiative across the whole region of Flanders. In 2015, 72 Eco Communal Gardens initiatives were established in Flanders. In 2016, the Flemish government allocated €320,000 for the establishment of twenty-five new ones. The Flemish association Velt vzw for eco-friendly living and gardening is the main driver behind the establishment of these initiatives. They have local groups active in 130 cities in Flanders, 16,000 members and 1000 active volunteers that practice and promote more eco-friendly ways of living and gardening.

Internal replication is when one neighbourhood initiative inspires other neighbourhoods to do the same of something similar. In Genk for instance, the Compost School and Pass-on shop (doorgeefwinkel) initiative have been replicated from one district to others. Another is the Brighton & Hove Food Partnership, that aims to spark sustainable community food initiatives across the city region of Brighton. Between 2009-2013, they expanded the number of communal food production projects in the city from 25 to 75 by providing space and support to local actors. They are now regarded as one of the leaders in the UK-wide network of Sustainable Food City Initiatives offering good practice for other cities.

Of course we are inspired by other initiatives. We transfer the experiences we make as members in other initiatives to our own initiative and the other way around.

Sören Rogoll, Transition Town Dresden

Fast diffusion of alternatives from city to city, is enabled by networks, open source knowledge, and coaching and mentoring programmes. For example, the Brighton (renewable) Energy Coop was inspired by an already existing community energy group in Oxford that provided mentoring to the Brighton Cooperative. In fact, one-to-one mentoring schemes – where people who have set up local renewable energy co-ops pass on their knowledge to help others do the same – are frequently occurring throughout the UK. Some of them are even funded through foundations or Co-ops UK. After receiving mentorship from Oxford, the Brighton Energy Coop is now also actively involved in helping others (e.g. in London). Another example of mentoring is Stockholm's Green Wedges Collaboration where the change makers behind the first initiative are offering support to new Green Wedge initiatives in other parts of the city that aim to do the same.

How transition initiatives are partnering to advance sustainability locally

The joint organisation of activities, events or projects is a popular way of promoting synergies. A striking example hereof is the yearly pumpkin contest that the Compost Masters of Genk organise to attract more people to the compost demo sites, since pumpkins ideally grow on compost. By partnering with Flemish, Turkish, Italian, Moroccan, Polish, Spanish, Slovenian and other women's organisations to organise a tasting market around the wide diversity of pumpkin recipes, they attract a multicultural group of visitors to the demo sites and make connections across the wide range of ethnic backgrounds of the residents of Genk. The Compost Masters thus smartly use food as a connector and a demonstrator to showcase that diversity enriches the lives of citizens and that composting is part of food production.

Partnering also promotes alliance building. For instance the Brighton & Hove Food Partnership connects local initiatives – such as urban gardens, foodbanks, cooking projects and community cafés – in a network. These small-scale initiatives are less able to engage in city-level debates by themselves. However, through regular engagement with the Food Partnership, many of their concerns and issues have a means to be vocalized. Through the Food Partnership multiple small initiatives can subsequently speak with a louder voice. The Szatyor Association in Budapest is another example of a local, citizen driven, organic food cooperative that brings together many family-scale producers and provides advocacy on their behalf. Alliances have more critical mass and are better suited to align interests and benefits across a multitude of partners. This also means that they have a better capacity to exert their strategic agency in a multi-level governance setting. Ultimately, partnering can give rise to complete new innovation networks such as the Living Coast initiative in Brighton that gave rise to a partnership of forty local public, private and civil society actors with a mission to pioneer a positive future that connects people and nature.

Some initiatives set up official or legally binding partnerships such as the community-supported agriculture farms Schellehof and dein Hof in Dresden in which participating citizens are both shareholder and stakeholder. Essential in partnerships is the building of trust which is the foundation for any long-term relationship. In Dresden, the urban garden network, that connects the different urban gardening initiatives in the city, was successful in setting up a cooperative arrangement with the public administration and by doing so, they managed to get urban agriculture on the political agenda and are about to incorporate it into the local spatial planning processes. The Heempark public-civic initiative was responsible for the implementation of the Bee Plan of the city of Genk and facilitates collaboration between civil servants and volunteers.

In the five city regions, most partnerships are established predominantly within a sector or low carbon domain. Examples include the Urban Agriculture Network Stockholm (Stadsodling Stockholm) that promotes networking between urban gardeners and matching between people and spaces or the Sustainable Business Partnership in Brighton that connects small businesses around sustainability topics to promote the sharing of knowledge and experience. We observed only a few initiatives that were able to build cross-sector partnerships like the Living Coast initiative in Brighton. This might indicate that the urban selection environments favour initiatives that mirror the way governments are organised, e.g. in separate departments such as a nature department, agriculture department, economic department, social department etc. While mirroring governmental structures is probably the most successful way for initiatives to get things moving locally, this also poses a severe problem of locking in unsustainability, as solutions often transcend departments, sectors and domains.

Several transition initiatives are aware of this shortcoming and are actively exploring how they can give rise to more systemic solutions that take into account the impact across sectors and domains. Exemplars of systemic thinking are transition town and permaculture initiatives. These initiatives crosscut domains and sectors and work towards more integrative visions and innovation pathways. Transition town initiatives actively explore how to give rise to a post carbon society through co-creation and experimentation on multiple domains simultaneously such as water, energy, food, housing, nature, education etc. Permaculture initiatives on the other hand explore how to design and maintain food production systems that work in harmony with nature and thus bring together food production, nature restoration and community building in one initiative. Such cross-cutting, systemic initiatives often have no one to speak to directly at the city administration and they thus have to negotiate with several departments at once, each with different visions and ambitions. This significantly slows down their development and thus hinders the positive impact they can generate for their urban environments.

> We linked up with other sympathetic kinds of organisations because you're a stronger force if you do. Loads of us are campaigning on environmental issues but coming at it from different angles. We are more powerful if we are part of a movement.
>
> Duncan Blinkhorn, Bike Hub Brighton

© Raf Gorissen

However, such initiatives are pivotal for scaling sustainable change since they are promoting systemic innovation by targeting the logic of value creation and by redesigning value chains. And this is not only important in terms of sustainability but also in terms of local resilience towards climate change and economic disruption because ultimately, they are giving rise to local economies. Partnering, especially between unusual partners such as the bicycle cooperative and organic vegetable farm in Budapest or the public, private and civic sector in the Brighton Living Coast initiative, thus contribute to the scaling of positive systemic change in cities.

How transition initiatives are upscaling more sustainable ways of thinking, doing and organising

Many transition initiatives in the five city regions have grown substantially over the past years. In Brighton for example, the Energy Coop started out as an idea of three people in 2010 and has since developed into the largest solar PV installer in the city with around 240 (mainly local) shareholders. In terms of monetary resources, the initiative grew over just six years from an initial investment of £18,000 by six people to having attracted over £1.5m in community shares. The consumer cooperative for regional and ecological food in Dresden, called VG (Verbrauchergemeinschaft für umweltgerecht erzeugte Produkte), started with just a handful of active members in 1991 and is now one of the biggest consumer cooperatives throughout Germany with approximately 10,000 members. Similarly, the Brighton & Hove Food Partnership grew from 200 members in 2008 to more than 4000 in 2016 due to a conscious push to increase membership size. In Genk, most initiatives did not record growth rates in number of members or visitors systematically, but sector reports of the public waste agency OVAM (2013) and the umbrella association of the re-use centers KOMOSIE (2016) for the wider region of Flanders show a considerable increase in sustainable practices related to a circular economy. For instance, home composting in Flanders increased from 41% in 2006 to 52% in 2012, while product re-use increased with approximately 6% from 2013 to 2014. This may not sound spectacular, but Flanders is already one of the top regions when it comes to recycling and re-use. For instance, in 2016, all the re-use centers of the region of Flanders attracted 5.8 million customers (of a population of about 6.4 million), re-used more than 32,355 tonnes of goods and gave 5426 people a meaningful job. Both the Compost Masters and re-use centres have been instrumental in promoting this increase.

While many initiatives are growing, for others the main focus is rather on sustenance than growth, indicating that the conditions for enabling growth vary across domains and locations. But when significant growth occurs, positive change towards higher levels of sustainability can quickly cascade to the wider system as the Sustainable Food Cooperative VG in Dresden, the Brighton & Hove Food Cooperative and the Szatyor Association in Budapest show. These examples show that one initiative can catalyse change across the value chain and beyond the city, cascading transformative processes across the country. The urban Bee Plan in Genk is an example where cascading positive change to other sectors is specifically targeted. Not only does the plan improve living conditions for bees on public land, it also involves the private and civic sector to do the same. This induced new bee-friendly civic initiatives as well as initiatives from the beekeeper association and the catering sector.

Upscaling to an impressive critical mass is also an effective method to weigh in on public policy-making. An upturn in the visibility of the Budapest bicycle culture through group riding events and an annual Critical Mass bike ride, attracting tens of thousands of cyclists, has brought bicycle mobility to the forefront of urban transport discussions in the city, which have led to direct updates in infrastructure in a number of districts in the city. The Hungarian Cyclists' Association behind the initiative, is the most significant civil organisation in the Hungarian cycling society and an affiliated organisation of the European Cyclists' Federation. Their member numbers are continuously increasing. The Association has built up a network of local activist bodies all around the country, and it organises volunteer professionals into working groups for infrastructure, communication and education, thus creating an effective matrix of local and professional knowledge in every field related to cycling. They succeeded in establishing enough professional credibility to be regularly included in scenario research and workshops organised by the Budapest Centre for Mass Transit.

Interestingly, for the majority of initiatives in the five city regions, upscaling is about increasing positive impact and thus not only about numbers. For most, sticking to their core principles and values is priority and many look for alternative ways to grow, such as replication, when they hit their maximum growth. Initiatives that favour replication to upscaling in size include Stockholm's Green Wedge Collaboration, Cargonomia in Budapest and the Pass-on shop in Genk. This illustrates that transition initiatives adopt another logic of value creation compared to business-as-usual. This also makes them very fragile when it comes to surviving the current economic selection environment. While the current value regime endorses economies of scale to reduce production costs and increase profit, transition initiatives focus on offering quality rather than quantity, on localized, responsible value chains and on multiple, collective and shared value creation rather than a sole focus on financial profit. They thus not only offer a countermovement against the still dominating economic logic that bigger is better, they are setting up new localized economies, where value creation and place are much more intertwined, advancing the logic of value creation.

How transition initiatives are instrumentalising windows of opportunity to fast forward sustainability locally

Transition initiatives successfully use the space and opportunities that are created by advanced national, regional and local policies. For instance, the 'Renewable Energy Act' in Germany has been a key factor in increasing the share of renewable energy production across the country. What is more, the energy transition in Germany (Energiewende) is driven by citizens and communities: Germans want clean energy and they want to produce it themselves. Together, the Act, the high level of investment security and the lack of excessive bureaucracy created favourable conditions for civil energy cooperatives to sprout. In 2013, nearly half of the investments in renewable energy production were made by small investors, providing communities with new jobs and increased tax revenues. Likewise, in the UK, a feed-in tariff scheme for renewable energy similar to that of Germany, significantly incentivized the deployment of solar energy installations while enabling citizen participation such as in the Brighton Energy Co-op. However, when the Renewable Energy Act was amended in 2011 and the feed-in tariff was reduced, solar PV installations became less profitable and harder to finance, which resulted in a significant decline in the number of new projects.

Lack of continuity (changing the rules of the game too quickly for initiatives to adapt) as illustrated above, hampers the scaling of positive change, as well as lack of transparency (hindering initiatives to seize opportunities), too much complexity (excessive bureaucracy) and still functioning but outdated policies and structures. For example, lack of governance transparency, of efficient policy support structures and accessible funding mechanisms in Budapest hinder initiatives from mobilizing the resources needed to advance local sustainability significantly. What is more, a lack of transparency enforces a lack of trust which makes initiatives hesitant in collaborating with the city administration. While this hampers scaling, on the positive side, this also fuels resourcefulness and creativity of local change makers, giving rise to highly innovative initiatives shaping new value chains. For instance, in the case of Cargonomia, a cooperation brings together bike welders, a group of concerned citizens, three organic farms and a socially conscious bike messenger company to simultaneously deliver sustainable services relating to logistics, communication, delivery and food.

Instead of seizing the opportunities, as illustrated in the German and UK example above, initiatives can also create new windows of opportunity. An inspiring example hereof is the Critical Mass bike movement in Budapest. While the government in Budapest for years failed to incorporate more environmentally friendly transport opportunities into their policy plans, a small group of engaged bikers started to collaborate and organised peaceful bike protests where a bunch of bikers take over a street and block car traffic for a short period of time. In just a couple of years, the initiative grew into huge Critical Mass bicycle rides that take over entire

To make the biggest wholesale change in what happens, you have got to take lots of people with you.

Vic Borrill, Brighton & Hove Food Partnership

streets to demand a more bike-friendly city. With tens of thousands of participants, the Critical Mass rides made it to the pages of foreign newspapers and attracted the attention of local politicians. The rides became festivals and cycling became appreciated and safer as the number of everyday cyclists grew bigger and bigger, illustrating that civil society, when organised well and collaborative, can create windows of opportunity for positive change even in the absence of supporting policies. Growing enthusiasm and a daily increase in the number of bikers on the streets of the city have encouraged a working relationship between the Budapest Mass Transport Organization (BKK) and the Hungarian Cyclist's Association (MKK). So instrumentalising, via the creation of windows of opportunity, can in effect lead to the embedding of more sustainable ways in policy plans and practice.

Transition initiatives can also effectively tap into opportunities at higher governance levels to promote sustainability locally, such as European funding schemes or global trends. For example, the organic food and gardening initiatives in Stockholm, Dresden and Genk have increased their proficiency, outreach and scale thanks to EU funding. EU funding schemes have also helped cities to create or renovate spaces as incubators for sustainable ways of thinking and doing, such as the Environment Centre in Dresden or the re-use centre in Genk. What is more, European funding is often thought of as more neutral, bringing together unusual partners in a setting where they can explore collaborative arrangements without the burden of dominating vested interests or political agendas influencing the possibilities.

Next to mobilizing funds, local initiatives can also mobilize new kinds of thinking and doing to their cities to get the ball rolling locally by tapping into global trends and networks. For instance, while the European Common Agricultural Policy is still biased towards large-scale, conventional agricultural businesses and, conversely, disadvantages small-scale, organic farms and urban agriculture initiatives, a bottom-up movement of sustainable agricultural practices is taking shape and inspiring many urban agriculture and gardening initiatives around the world. They tap into new design trends coined as 'living systems thinking, bio-inspired innovation, permaculture, regenerative design and development' – learning from the living systems in the natural world and how they are organised to inspire human design.

One of those new design approaches pioneering more sophisticated life-friendly food production practices that inspired initiatives in the five city regions is permaculture. Founded in the 1970s, this new approach sparked meaningful action by self-organised groups of concerned citizens in every corner of the globe. Permaculture is an exemplar of working *with*, rather than *against* nature, designing regenerative food production systems, human habitats and organisational arrangements that work in harmony with nature. Transition town initiatives as well as urban agriculture initiatives are inspired by these permaculture principles to develop productive, healthy, resilient, life-friendly food production systems. The network offering teaching and exchanging experiences about permaculture is growing and spreading from the bottom up giving rise to a global grassroots web of professionals, practitioners, educators and new enterprises. Tapping into this grassroots network to learn new skills is a resourceful way of building local capacity for more systemic solutions. National platforms, such as the Dutch-Belgian Permaculture School, the Academy of Permaculture in Germany or the UK Permaculture Association, are important network nodes or hubs to promote this more systemic kind of thinking, designing and practicing.

How transition initiatives are embedding sustainability in the local governance

When it comes to changing ingrained unsustainable practice, transition initiatives are instrumental and influential in inducing behavioural change in citizens. First of all, citizens involved in these initiatives bring into practice more sustainable ways of thinking, doing and organising and thus lead by example. By demonstrating eco-friendly alternatives in demo sites, educational activities, events, festivals etc., transition initiatives showcase that alternatives work in practice, they advance collective learning and diffuse alternative ways to the wider public. A good example hereof is the Velt initiative in Flanders of which the local Genk branch maintains various demo sites and organises a multitude of events and activities promoting an eco-friendly lifestyle. Another great example is the Waste House in Brighton showcasing that what is generally considered as waste can also be a valuable resource for something new, while the development process involved a multitude of actors ranging from students to building contractors. Both Magnet Bank in Budapest and the Waste House in Brighton also serve as spaces for community activities.

When it comes to changing consumption patterns, transition initiatives are the key innovators that offer more sustainable products and services to citizens. The VG shops in Dresden for example offer environmentally friendly products to citizens and the energy co-ops in Dresden and Brighton offer renewable energy to civic co-investors. Other examples include the numerous food initiatives in Dresden, Stockholm, Genk, Budapest and Brighton that show that food can be grown in the city in an environmentally friendly way. Furthermore, urban agriculture initiatives promote the reconnection between citizens, food production and nature. This impact can be greatly reinforced when it is supported by governments. For instance, the municipal organic food initiative in Södertälje, Stockholm, contributes greatly to the mainstreaming of sustainable and eco-friendly food by only serving healthy, seasonal and locally produced organic food. The Climate Protection Unit (Klimaschutztab) of Dresden on the other hand is an important catalyst for climate-friendly mobility, energy efficiency and renewable energy in the city. The district budget in Genk specifically seeds bottom-up initiatives that contribute to the

We create networks of values, that in turn create structures.
If you change values you can change the physical structure.

Bette Lundh Malmros, Stockholm County Council

local community like the Pass-on shop (Het halve eurootje) that promotes product re-use and offers inexpensive clothing to citizens less well off.

Changing practices requires changing mindsets first and several initiatives explicitly focus on building dialogues to increase citizen awareness and environmental consciousness. Examples include the Brighton Peace and Environment Centre which organises local 'Carbon Conversations' for citizens to discuss environmental issues in a deeper way. The Werk.Statt.Laden recycling and repair initiative in Dresden regularly organises Neighbourhood Talks to reflect upon the development of the neighbourhood. The business led initiative Sustainable Hökarängen catalysed citizen dialogues to involve them in the decision-making processes and the civic initiative Transition Värmdö in Stockholm organises a yearly event called 'A week for the future' to stimulate locally based dialogues about eco-friendly solutions which attracted already 2000 visitors. This event is now replicated to a multitude of other Swedish cities to spark engagement and increase knowledge on sustainability.

Embedding more sustainable practices can also be promoted via creative activism, as the Critical Mass bike rides in Budapest illustrate. By doing this they both promote a low carbon transport mode and get more sustainable transport on the political agenda. Another interesting example in this regard is the Bee Plan in the city of Genk. The Bee Team, a group of thirty-five local volunteers, is active at many local events, often dressed up as bees, raising awareness around the fact that urban environments need to become more bee-friendly. They inspire citizens to think about this topic via creative actions such as handing out bee-friendly seeds or plants for citizens to plant in their gardens. Years before, Velt Genk, the association for the promotion of eco-friendly lifestyles, successfully campaigned to ban harmful pesticides from public green management via their 'No chemicals, better health' campaign.

Even more interesting is that many transition initiatives bring together the 'social' and the 'environmental' in their initiatives and often work promoting both social cohesion and environmental sustainability in their communities. Examples range from engaging deprived groups of citizens in food related activities, providing those without gardens with communal land to grow their own food, uniting citizens of all backgrounds in allotment gardens, to developing food gardens in school playgrounds. Initiatives are often very resourceful and creative in developing novel tools or guidelines to help citizens acquire new skills and practices. One of the Compost Masters in the city of Genk for example, developed a visual guideline about composting for children that have problems with reading. The Humusz Alliance of Budapest, a local organisation promoting zero-waste living, developed a summary package of composting requirements for schools and housing collectives, in addition to securing resources for promoting and establishing composting spaces in local schools. The Pass-on shop and re-use centre in Genk, like the

Refo initiative in Stockholm, promote product re-use while offering goods at affordable prices for low-income households. Refo and the re-use centre also employ disadvantaged personnel that have difficulties finding a job. The transition town initiatives Transition Wekerle, Dresden im Wandel and Transition Värmdö enroll community building activities around a wide variety of low carbon topics.

Another indispensable way of mainstreaming sustainable alternatives is via institutional innovation, because advancing positive impact is most probable when cities anchor sustainability into the heart of local city making and commit and invest in a guided process of institutional innovation. Interviewees of the city of Brighton for example, point out that their city offers a fertile breeding ground for setting up sustainability initiatives as there is a community of like-minded people and supportive local councillors and MPs who take sustainability seriously. In addition, the existence of a dedicated Sustainability team within the local government – and its willingness to give substantial support, especially time – has been helpful across a wide range of transition initiatives, including the Brighton & Hove Food Partnership, the Green Centre and the Hanover Action for Sustainable Living initiative. Even though the One Planet Living standard that Brighton adopted as a framework is still far from achieving its goal, it did help to move sustainability out of the margins of decision-making into the heart of the city-making process.

While sustainability has not yet entered the heart of the city-making process in Genk, the city can be considered as a frontrunner in shaping institutional innovation. By organising master classes for civil servants about systems and transition thinking, new perspectives are introduced to the city staff that promote reflexive thinking, and lay the groundwork for developing new skills and new roles. City leaders acknowledge thereby that to change the city, the city needs to change its thinking. For Genk, empowering citizens and involving them in policymaking is a priority and the city has organised several large civilian brainstorms to inspire local policy programs. District budgets and district managers help to align top-down and bottom-up activities. Sometimes, just one change maker in the right place can prompt the embedding of more sustainable ways into formal governance institutions. For instance, in Dresden, a pioneering governmental official from the Office for Urban Forestry and Waste Management opened up space for several transition initiatives to further develop their activities despite legal constrains. Especially the friendly collaborative environment that was created between the Office and the initiatives promoting urban agriculture, fuelled mutual benefit creation for both the volunteers and the government.

What is more, institutional innovation, if organised well, can be quickly embedded in city governance, as illustrated by the Bee Plan in Genk. Two citizens, concerned with the wellbeing of bee populations, showed the international documentary *More than Honey* during an open

session of the environmental council in May 2013. Afterwards, the audience brainstormed how their city could become bee-friendly. The ideas quickly developed into an encompassing Bee Action Plan that was approved by the bench of aldermen in 2014. The city provided the necessary funds and the volunteers of the Bee Team provided the time needed to implement the actions on the ground and improve the living conditions for bees in the city, illustrating that a constructive, collaborative governance context can accelerate the pace of positive change considerably.

In fact, as most cases illustrate, mainstreaming sustainable alternatives depends to a large extent on successful cooperation. The Brighton & Hove Food Partnership focused from the start on connecting different food actors and initiatives in a constructive setting while also building a collaborative relationship with the city council. That way they succeeded in developing and establishing the first UK city food strategy that is now being copied to other pioneering cities. The initiators of Critical Mass Budapest brought together all kinds of bicycle lovers and got them to collaborate to organise street take-overs so impressively that they convinced the ruling government to invest in making the city more bike-friendly. The Heempark in Genk drastically increased its outreach and activities when the volunteers and government set up a public-civic partnership. The Greater Stockholm Nature Guides initiative took off thanks to a collaboration between County employees and volunteers. The citizen cooperative Burgercraftwerke in Dresden collaborates with a multitude of stakeholders and municipal actors and this is just a selection of the examples of collaboration we came across.

© KPC Genk

PART THREE:

TOWARDS FUTURE-FIT CITIES

Chapter 6:

Key insights

The best life insurance for any species in an ecosystem is to contribute usefully to sustaining the lives of other species, a lesson we are only beginning to learn as humans.

Elisabet Sahtouris, Evolutionary Biologist

Sustainable change in cities requires active involvement of citizens. Policy makers need to trust the knowledge, experience and creativity of its peoples. When civil servants, artists, scientists, policy makers and entrepreneurs join forces, they co-create a different society. Wetopia, as I like to call this, is already under construction.

Joke Quintens, Former First Alderman of Genk

Meaningful change is happening organically, everywhere, all the time

While the insights and learnings from transition initiatives and how the mechanisms of acceleration play out are interesting in themselves, it gets really exciting when you zoom out. Looking at the dynamics of change from a bird's-eye perspective shows that meaningful change is happening organically, everywhere, all the time. At the heart of this change are urban change makers, civic innovators that invest in a more culturally rich, socially just and environmentally friendly future. Their transition initiatives are incredibly diverse, place-based, self-organised and collaborative. What is more, they are generally ethically-grounded, holistically-oriented and grounded in open-ended collective learning processes. To fastforward urban sustainability these initiatives set out to evolve the ways of thinking, doing and organising in their city. To achieve that, they use a wide variety of approaches that we classified as replicating, partnering, upscaling, instrumentalising and embedding. And their impact is not marginal, a recent publication of the ECOLISE network, a European network for community-led initiatives on climate change and sustainability, highlighted that in 2016 there were 3 million permaculture practitioners in 140 countries, 15,000 ecovillage communities on 6 continents and 1200 transition town initiatives in 48 countries. The oldest transition town initiative in Totnes, England, has been reported to involve 468 households that each save at least 1.3 tons of carbon emissions annually, while studies on carbon emissions in Danish ecovillages find they are 60% lower than the national average[11].

Permaculture initiatives not only draw down carbon from the atmosphere and sequester these in soils and plants, regenerative agriculture practices also avoid carbon emissions in the first place by using more sustainable methods like no till, no pesticides or synthetic fertilizers. Regenerative agriculture in fact is one of the most effective and promising avenues to reverse climate change while producing food without the negative side effects of conventional agriculture. Project drawndown estimates that a tenfold increase in adoption could result in a total reduction of 23 billion tons of carbon dioxide from both sequestration and reduced emissions[12]. These change makers cannot do this alone, however. They depend upon authorities to create the right conditions for meaningful and beneficial change to scale. City authorities are uniquely placed to reimagine the way their cities inhabit this planet and how they can contribute to the rest of life. Instead of managing cities at the expense of our life support system, city leaders can build regenerative capacity into their cities by inviting higher levels of consciousness, capability, creativity and co-responsibility. This goes beyond lighthouse projects. It is about cultivating the ability for rapid and collective learning and about establishing mutually benefical relationships between urban communities and the environment that increase the viability and vitality of both. Future-fit urban leadership is about calling forward a collective vocation for the city, facilitating regenerative value adding processes and integrating efforts into collaborative innovation networks that challenge the status quo[13].

Change starts inside

All studied transition initiatives start from the purpose of making a meaningful contribution to the future of their city: contributing to a circular economy, stimulating low carbon mobility, investing in a renewable energy system, increasing food sovereignty, restoring ecosystems, promoting social cohesion, regenerating local economies, renaturing citizens and city planning efforts. So while many dominant institutions are invested in the past, transition initiatives are creating the future. Even if this means going against the mainstream dynamics. Their bold and spirited undertakings address a largely neglected dimension of transitions: the socio-spiritual one. For the social, ecological and economic crises upon us are not just a crisis of systems – they are a crisis of spirit too[14]. If we genuinely want to transform our cities, we will have to evolve our thinking and our way of being. A first step in this regard is to become aware of the assumptions, beliefs and values that kept the unsustainable system in place in the first place. One of these is the illusion of separateness between humans and nature which is still deeply entrenched in our mindsets. This is painfully obvious in the dominant eco-efficiency discourses that still largely focus on hardware solutions, coercing the view that sustainability is mostly about technology. These discourses overlook the relevance of changing our mental software that runs the hardware. In other words, to transition our systems, we will need to transform our mental models. And many urban transition initiatives are working to affect such change by increasing awareness, by building eco-literacy, by promoting collective learning and by advancing consciousness so that citizens become more aware of their interdependency with the rest of life. So instead of growing our economy, we should be focusing on growing ourselves as benevolent beings first.

Food is a promoter and unifier for urban sustainability

Across all five city regions under study, food appears to be the domain in which transformative change is happening the fastest. Just think of all those networks like transition towns, community-supported agriculture and permaculture that are multiplying sustainable food practices at dazzling speed all over the globe. Food is clearly a unifier. Not only do food-oriented initiatives bring a wide diversity of people with different backgrounds and cultural roots together, like the Eco Communal Gardens in Genk or the Brighton & Hove Food Partnership, they also inspire new and more inclusive governance models like the city food strategy in Brighton, new value chains like Cargonomia in Budapest or new value logics like the Municipal organic food initiative in Södertälje, Stockholm. Most food-oriented initiatives also bring people closer to nature in a variety of ways. For instance, the Heempark in Genk engages citizens in a wide range of human-nature reconnecting activities while citizens become co-investors in healthy

> **I invite you to slow down and reconnect with yourself as well as your fellow human beings. Because real transition can only start within you and me.**
>
> Julia Leuterer, Dresden

Feed them lovely food and they will love you forever.
Jess Crocker, Brighton and Hove Food Partnership

and eco-friendly food production in the community-supported agriculture initiative Schellehof in Dresden. This is a significant insight, because a recent largescale study led by Paul Hawken[15] on the most substantive solutions to reverse climate change, demonstrated that rethinking the way we use, consume and produce food is among the most impactful solutions. Our findings illustrate that urban change makers are already full on working on these solutions. Food is thus a promising domain to accelerate fundamental change in a culturally rich and inclusive way. What is more, our results also show that the innovators for sustainability are often outsiders, citizens taking responsibility for fundamentally different ways of food production instead of incumbents from the food sector. This highlights that advancing urban sustainability is about investments in the new system and divestments from incumbent regimes that are no longer fit for the future by empowering the outsiders and planning for and facilitating disruption of the unsustainable system.

Sequencing change by working on new patterns

Another noteworthy observation when you zoom out is that there are central themes or unifying patterns across the workings of transition initiatives in our five study areas. These represent the ways in which change makers are rebooting urban living in their cities through their initiatives. Most of the transition initiatives work on reimagining, reconnecting, renaturing, repurposing, reorganising, reskilling and redistributing. They thus show that advancing sustainability is about verbs, not solutions. It is about changing the patterns of thinking, value creating, organising and collaborating. It is about establishing new relationships and value adding roles that can cascade beneficial change throughout the urban ecosystem. Solutions are always context-dependent – what works in one city is no guarantee to work in another because there the conditions might be different. So while solutions can be considered as 'fixed formula', working on shifting the patterns can be tailored to many different situations. They can serve as guiding templates. Such fine-tuning is indispensable in bringing forth genuine and enduring solutions that are place-based and locally attuned. From this exploratory study, we could identify at least ten things to know and do to fast forward urban sustainability. This list is by no means complete, nor is our understanding of them, it is work and knowledge in progress. But they do highlight the critical groundworks for accelerating urban sustainability.

Chapter 7:

Ten things to know and do to fast forward urban sustainability

Without order nothing can exist –
without chaos nothing can evolve.

Oscar Wilde, writer

TEN THINGS TO KNOW AND DO TO
FAST FORWARD URBAN SUSTAINABILITY

#1.
The next big thing will be a lot of small things

STRATEGY 1
DRASTICALLY INCREASE THE INVESTMENTS IN THE GERMINATION AND GROWTH OF MANY SMALL SCALE TRANSFORMATIVE INITIATIVES AND ORGANIZE THEM IN COMMUNITIES OF PRACTICE AND LEARNING NETWORKS

If you think small-scale, local transformative initiatives are just fiddling about, think again. A quick mapping exercise in the five city regions revealed about five hundred initiatives that in one way or another focus on environmental sustainability. At first glance, their transformative capacity might be tiny in a world dominated by large corporations and powerful vested interests. They are however not alone: the collective impact they bring about is not negligible. We see that transformation is spreading organically, in multiplicity, everywhere and that small transformative initiatives replicate amazingly fast around the globe.

In just a decade the transition town initiative spread virally to 48 countries, involving hundreds of thousands of households in creating low carbon, resilient and environmentally friendly communities. The same holds true for permaculture and eco-village initiatives that quickly developed into a movement engaging millions of citizens worldwide that are changing the ways we produce and consume along the way. In just a couple of years, the first repair café initiative, organised in the Netherlands in 2009, grew to approximately 750 initiatives that spread to Japan, US, Canada, Australia and many more countries. Urban gardening initiatives are back on popular demand and energy co-ops, re-use centres and renaturing city initiatives are replicating at a high pace across the globe. The systemic power of transition initiatives lies in their numbers. Instead of the dominant 'too big to fail' growth paradigm of the corporate world, their progression is about 'being too many to ignore'.

Our research shows that well-organised initiatives, despite their relatively small size, can influence the wider system beyond the city boundaries via positive spillover effects, outward replication or via the reshaping of the local governance context so that it is more open for business-beyond-the-usual. What is more, urban change makers are not only multiplying change, they are diversifying change by tailoring alternative ways of thinking, doing and organising to fit their place. Viewed from a transition's perspective, local transition initiatives are giving rise to place-based regenerative solutions as a counter movement to globalizing uniformity. Many tiny transformations will aggregate into big change. The next big thing will be a lot of small things.

ACTIVITIES TO ORGANISE

Co-creation	*Envisioning*	*Dialogue space*
Crowd sourcing	*Participatory budgeting*	

Ten years ago cycling was not popular in Budapest at all. It was viewed as a sport or subculture activity and not a means of transport. Thanks to the 'I Bike Budapest' mass bike rides, attracting tens of thousands of cyclists, the Budapest bike movement is back and bigger than ever.

Aron Halász | Budapest

TEN THINGS TO KNOW AND DO TO
FAST FORWARD URBAN SUSTAINABILITY

#2.
The city of the future is designed for life (not cars)

STRATEGY 2
RETHINK AND REDESIGN CITIES FOR LIVEABILITY, VITALITY AND VIABILITY BY ADOPTING 'LIFE-FRIENDLY' AND 'LIFE-ENHANCING' AS CENTRAL QUALITIES IN URBAN PLANNING PROCESSES

Many of the studied initiatives in Stockholm, Budapest, Brighton, Dresden and Genk aim to bring nature back into the city or reclaim space and acknowledgement for the living world. From the Green Wedge collaboration in Stockholm, urban beekeeping in Genk, urban gardening in Budapest, community-supported agriculture in Dresden to the Biosphere Partnership in Brighton, these initiatives all aim to either renature urban environments or safeguard natural environments from human development. The fact that human versus natural environments are not necessary contrasting opposites is beautifully illustrated in the Children's Psychiatric Centre (KPC) in Genk. To create a healing environment, architects integrated the human and natural world by sinking the building into its environment. By weaving therapy rooms and gardens together in a semi-natural landscape, the building promotes the wellbeing of patients and employees while offering habitats and corridors for plants and wildlife.

Some cities take it a step further, like Adelaide in Australia, which is creating an urban forest comprised by millions of trees and shrubs or the Chinese Liuzhou Forest city that will be entirely covered by plants and trees. The fact that people want to bring nature back into the city is not surprising. We have largely designed our urban habitats as concrete deserts. Most space in cities is dead space – space for driving and parking cars and storing goods. Such space is not life-friendly – neither for humans nor for other forms of life. It leaves citizens craving for healthy air, better liveability, quiet refuges and spiritual, recreational and aesthetic landscape values. Many change makers thus devote their time to bring nature back to the city.

Redesigning cities for life requires a shift from technical to living systems design. This can only be achieved if humans acknowledge their interconnection and interdependency with all other living things and anchor 'life-friendly' and 'life-enhancing' as central qualities in urban planning processes. This means that renaturing cities is much more encompassing than the development of engineering solutions for green infrastructure such as green walls or roofs. Designing cities for life requires a whole new set of (re-)design principles and a complete shift from reductive thinking to relational, pattern and systems thinking as well as shifting from degenerative to regenerative value logics. Only when we accomplish to make this shift, we can start to develop cities that support life over the long haul.

DISCIPLINES TO EXPLORE

Biomimicry	*Biophilic design*	*Nature-based solutions*
Bio-inspired innovation	*Living systems thinking*	

> Sustainable development needs public urban space; not just public aisles between privatised bunkers.
>
> Jacqueline Muth, Dresden City Council

TEN THINGS TO KNOW AND DO TO
FAST FORWARD URBAN SUSTAINABILITY

#3.

Human – nature – human nature reconnecting is key

STRATEGY 3
RENATURE HUMAN NATURE BY ENCOURAGING NATURE EXPOSURE AND BY NURTURING ECO-LITERACY, SYSTEMIC CONSCIOUSNESS AND PERSONAL DEVELOPMENT

In line with the previous insight, many of the identified initiatives are also explicitly working towards reconnecting citizens to nature, to each other and to human nature. By doing this they fill a void left by the incumbent institutions that mainly promote discourses of eco-efficiency and technological fixes. While technology is important, it is only part of the solution. As Albert Einstein pointedly stated: "You cannot solve the problems of today by using the same thinking that created them." This is exactly where technological solutions run short, for neither eco-efficiency nor technological fixes get to the root of the problem: the illusion of separation between humans and nature. Nature is not just what is out there – we are nature too. We will not be able to resolve problems of unsustainability as long as our mental models promote anthropocentrism – humans on top of the pyramid, instead of biocentrism – humans as part of the web of life. Increasing urbanization as we know it (without considering nature), and the way nature is mostly designed out of the cities, however, further increases an alienation from nature, reinforcing the cause of our current crises.

Reconnecting humans to nature
Humanity is out of sync with the rest of the world. Our lack of awareness that human life is interconnected and interdependent with the rest of life on this planet, is the main cause of overexploitation, pollution and deforestation. It is not a coincidence that we are burning ourselves up at an alarming rate in times of global warming. It is not a coincidence that our immune systems are eroding in places of environmental degradation. What we do to nature, we are doing to ourselves. What is more, alienation from the natural world or a lack of access to nature aggravates mental, behavioural and physical problems. Coined as 'nature deficit disorder', this topic is gaining more and more attention in the world of science. The fact that so many urban initiatives focus on the reconnection of humans with nature, implies that change makers are (consciously or unconsciously) aware of how this disconnect reinforces unsustainable ways of thinking, doing and organising. In a response, local change makers have developed a wide variety of initiatives to re-establish the connection between citizens and nature. They use diverse ways of bringing nature closer to citizens and often establish experiential learning initiatives involving the natural world: they bring nature to schools and communities via schoolyard

and communal gardening plots, they involve citizens in nature immersion and conservation activities, they experiment with growing food in harmony with nature, they organise compost schools and train nature guides and so much more.

Renaturing human nature

We are not only disconnected from nature, we have also lost connection to our own nature. While the current dominant value logic drives consumerism, individualism and competition, innate human tendencies get starved. We are a social species and therefore need each other to succeed[16]. Many urban initiatives are driven by a desire to promote social inclusion and cohesion. Initiatives such as the Brighton & Hove Food Partnership, the Eco Communal Gardens in Genk or the transition town initiatives from Stockholm, Dresden and Budapest explicitly adopt, develop and improve community building approaches to reconnect citizens in new ways. Several initiatives also explicitly incorporate strategies for personal development as they recognise that upscaling benevolent change is as much about personal innovation as it is about institutional innovation. Transition town initiatives for instance, have a specific focus on what they call the 'inner transition' and many initiatives promoting sustainable resource use, such as sharing platforms and tool libraries, stimulate people to think differently about objects and ownership. Permaculture is about designing regenerative ecosystems that provide food in harmony with nature. It thus explicitly develops systems thinking in practitioners. It is obvious that we cannot change our ways unless we change our thinking. Renaturing human nature is about building capacity for evolution, growing ourselves as human beings and changing our relationship with the rest of the living world.

INITIATIVES TO INVEST IN:

Urban agriculture	*Permaculture*	*Urban rewilding*
Food forests	*Forest bathing*	

The benefits of being outdoors and reconnecting to nature are limitless.
Miriam Steiner, Brighton

TEN THINGS TO KNOW AND DO TO
FAST FORWARD URBAN SUSTAINABILITY

#4.
Forget power or status. Urban change makers are motivated by passion and purpose, shifting leadership from ego-based to soul-based

STRATEGY 4
INVEST IN NEW STYLES OF DISTRIBUTED LEADERSHIP THAT ARE SERVICE-BASED AND REGENERATIVE AND EVALUATE LEADERSHIP IN TERMS OF HOW IT ENABLES THE VALUE GENERATING CAPACITY OF OTHERS

The majority of transition initiatives studied are purpose and impact driven rather than profit driven. Even though they have different missions, all of them are founded on the ambition to contribute to the society in a meaningful and beneficial way. Change makers are energized by passion and purposefulness. Intrinsic motivation is what fuels and drives them. Their leadership styles are service-oriented. They are thus deviating from contemporary leadership styles based on extrinsic motivation and driven by power and status. Their initiatives are about much more than profits. They are about inclusivity, resilience, harmony and other qualities of prosperity. They are about taking responsibility for the future. They aim to emancipate the systems in which they work by breaking through the status quo. Urban change makers often had to move mountains to get their initiative going since the current system is more often than not overly complex and unfavourable to issues related with sustainability. Perseverance and ingenuity are thus part of change makers' skillsets for creating pockets of the future in the present. Often, change makers have to make a tough decision to either 'fit and conform' to the current selection environment or 'stretch and transform'[17] current ways of thinking, doing and organising to achieve meaningful change in their cities.

If they conform to the current value regime they have a better chance of influencing city agendas and embedding more sustainable ways in the local governance practice. A 'fit and conform' strategy can therefore promote the mainstreaming of alternatives because it allows a quicker uptake, just think of the organic agriculture and re-use examples. These do not challenge ruling paradigms and are thus more easy to implement. Some initiatives rightfully resist to fit and conform to a system that is clearly inapt to sustain itself over the long haul. Instead, they stretch and transform dominant paradigms. Energy co-ops for example, enable civic decentralised power generation and public-civic partnerships, like the Heempark in Genk and the Green Wedge partnership in Sweden, pave the way for co-investment and co-responsibility between citizens and public authorities, transforming the rules of the game along the way.

Several initiatives across the five city regions also deliberately resist scaling in the old-fashioned hierarchical way, such as the Pass-on shop in Genk, the Brighton & Hove Food Partnership or the Szatyor Association in Budapest. In hierarchical models, the most powerful decision makers are furthest from the frontlines, lacking the local details required to make good decisions. In addition, the longer the chain of command, the slower the responsiveness, while communication failures mount. To overcome these shortcomings, many transition initiatives advocate growing in networks where leadership is shared and distributed instead of concentrating leadership and power at the top of a pyramid. These initiatives thus prefer to grow in reciprocal networks that can cross the boundaries of cities, nations and even continents and by doing this, they are pioneering new organisational development models that might even inform the business world. Future-proofing cities is thus about developing new leadership styles that can create the right enabling conditions for forwarding urban sustainability. These enabling conditions are about empowering, empathising, evoking a sense of purpose and a shared vocation, encouraging relational thinking, nurturing collective learning and identifying synergies.

CONCEPTS TO EXPLORE

Regenerative leadership *Self-organisation* *Mindfulness*
Positive psychology *Service-based leadership*

> Transitioneurs and activists are living the very life and reality we want to create, showing us that it can be done and it is not just a question of 'no, no, no', but rather a different 'yes, yes, yes'.
>
> Magdalena Frembäck | Stockholm

TEN THINGS TO KNOW AND DO TO
FAST FORWARD URBAN SUSTAINABILITY

#5.
Welcome urban hacking, to create the future is to hack the past

STRATEGY 5

EXPLORE NEW ROLES OF CONSCIOUS, VALUE ADDING CITIZENSHIP THAT CAN CASCADE BENEFICIAL CHANGE FROM NEIGHBOURHOODS TO COMMUNITIES TO REGIONS TO THE WORLD

Moving towards low carbon, life-friendly and equitable cities is about co-creating new urban governance models that are self-organising, responsive, adaptive, creative and resilient. Current systems however are rigid, outdated and overly organised. They are designed to repeat the past. Transforming such systems requires significant interventions. Many urban change makers are doing exactly that: they hack the system by enacting new value adding roles. If the system is no longer fit for purpose, someone needs to intervene to change the rules of the game. Such transformation can be initiated by any change maker, whether they are from civil society, business or public authorities and requires serious out-of-the-box thinking. It can even require creative insubordination or civil disobedience, like the Critical Mass bike rides in Budapest to urge policymakers to make the city more climate- and bike-friendly. But also city leaders play an important role in opening up space for fundamental innovation by ignoring, discarding or removing inappropriate rules.

To create a city that is fit for the future therefore requires systemic change and our research shows that changing roles is a potent way to hack the old system. For instance, renewable energy co-ops shift from being consumers to being producers (prosumers), public authorities like Genk are shifting from a controlling to a more facilitating style of management, permaculture initiatives are shifting agricultural practices from controlling to enabling nature, CSA initiatives are making consumers shareholders and many urban gardening initiatives are turning citizens into stewards of public open space. But with new roles come new responsibilities for businesses, citizens and governments. Governments for instance, will have to expand their current regulatory role with also becoming a deregulator, taking away excessive bureaucracy, unnecessary and unproductive rules and regulations that hamper sustainable innovations. For, accelerating sustainability requires a smart combination of divestment from the old system and investment in the new system. As incubators for radical innovation, governments can promote disruptive planning and experimentation to challenge the status quo and they can create the conditions for regenerative instead of degenerative value creation. Local businesses play an important role in showcasing that regenerative value creation 'doing well by doing good' makes business sense and they play an important role in promoting local economic resilience. Finally, citizens will have to step up to take co-responsibility over certain aspects of their city and stewardship over urban commons.

Most of all though, new roles need to be underpinned by new abilities and skills, which in turn need to be enabled by new mandates. To reconnect citizens to nature and awaken a shared urban vocation to regenerate life, cities need to empower systemic socio-ecological architects, people skilled in weaving nature-based solutions and social processes in a way that promotes the viability of both. To redesign planning and decision-making processes, cities need to enable path-breakers skilled in systemic innovation and co-creation. To enable partnerships and increase the quality of collaboration and co-operation, cities need to empower translators and connectors skilled in mediating, uniting and facilitating. To align, coordinate and aggregate scattered activities into something that is more than the sum of its parts, cities need to enable integrators skilled in relational, pattern and systems thinking. And each of these new roles requires a genuine mandate for implementation in the city system that is not obstructed by outdated fossilized structures and cultures or dominated by vested interests reinforcing the status quo.

Often forgotten but very important is that these new roles will only be successful if they go hand in hand with new tools: new guiding principles, new evaluation criteria, new urban planning and design guidelines, new democratic processes, new accounting models etc. You simply cannot post a tweet with a typewriter. These tools will have to be developed by the cities themselves. How is a city designed for life different from a city today? What characteristics are essential and how can they be measured? How can new value adding roles and new skills be developed and enabled and what does this mean for evaluation criteria? How can the logic of value creation be altered? Crucial here is to invest in a city-wide strategy to anchor and build capacity for 'transitioning' the city to higher levels of sustainability and resilience. As mentioned in the previous point, building the city of the future is about building capacity for evolution into urban governance. New solutions are seldom perfect from the start, so we will have to learn to better deal with imperfection and invest in fast but continuous pathways of improvement.

THINGS TO PROMOTE:

Open innovation	*Consciousness hacking*	*Open source*
Governance experimentation	*Ecosystem stewardship*	

TEN THINGS TO KNOW AND DO TO
FAST FORWARD URBAN SUSTAINABILITY

#6.
Forget competition. The city of the future is built on collaboration and powered by reciprocal networks

STRATEGY 6
FOSTER AND REWARD COLLABORATION AND BUILD CAPACITY FOR IMPROVING THE QUALITY OF RELATIONSHIP, ALLIANCE AND NETWORK BUILDING

Our research shows that collaboration catalyses the scaling of more sustainable practices, especially the cross-sector ones. From the Biosphere Partnership in Brighton, the Green Wedge collaboration in Stockholm, the VG initiative in Dresden, Cargonomia in Budapest to the Eco Communal Gardens initiatives in Genk, they all scaled their impact via collaborative arrangements. More often than not, these collaborations involve combinations of the public, private and civic actors. Even more impactful is when the government actively promotes collaboration while investing in the right empowering and enabling conditions. For instance, in Genk, district managers and community builders help catalyse mutual beneficial relationships between civil and policy actors to spur meaningful urban innovation processes. Collaborative and open innovation approaches even give rise to new governance models such as the Heempark public-civic partnership or new business alliances like the Cargonomia initiative in Budapest. Other examples include the Brighton & Hove Food Partnership that managed to develop the UK's first urban food strategy after partnering with the city administration, the Dresden Urban Gardening Network that managed to get urban agriculture on the city planning agenda by fruitfully working together with the local public authority or the Sustainable Hökarängen initiative that brought together a wide variety of citizens in the decision-making processes to decide the future of Hökarängen.

Scaling urban sustainability is thus not a matter of bottom up *versus* top down but of bottom up *with* top down. Our analysis shows that both bottom-up initiatives and the local government play an instrumental role in fuelling acceleration dynamics and either one can take the lead in reaching out to the other. Sometimes transition initiatives first have to openly challenge public authorities to become more ambitious in scaling local sustainability such as the bike movement in Budapest did with their Critical Mass rides. Sometimes the public administration needs to challenge its citizens to adopt more life-friendly behaviours like the Municipal Organic Food initiative in Södertälje or the urban Bee Plan in Genk illustrated. Sometimes, citizens and public administrations find a way to co-constitutively scale transformative innovation via collaborative public-civic partnerships, like the Heempark in Genk or the Energy Coop in Brighton. From a meta perspective, it appears that the pace of positive change is mostly prompted by new kinds of cross-cutting collaboration. For instance, the Waste House in Brighton – co-created by constructors and students – generates more and more projects related to sustainable living, the Eco Communal Gardens in Genk – co-created by public and civic actors – initiated a region-wide support strategy multiplying eco-friendly gardening initiatives across Flanders or the Green Wedges Collaboration in Stockholm where NGOs and municipalities together pave the way for a Green Wedge Council to safeguard the cities' natural capital. Creating enabling conditions for an open innovation environment that welcomes plurality of approaches, guarantees inclusivity and rewards cooperation will snowball collaboration and propel cities forward towards the future.

THEMES TO INVESTIGATE:

Community building
Personal development

Customer centricity
Impact investment

Regenerative design and development

> We are constantly growing;
> but is the growth concept the right one?
> How can we develop qualitatively,
> as well as in terms of quantitative upscaling?
>
> Florian Etterer, Urban Gardening Education Initiative Dresden

© Raf Gorissen

TEN THINGS TO KNOW AND DO TO
FAST FORWARD URBAN SUSTAINABILITY

#7.
Building a new logic of value creation is the only way out of the status quo

STRATEGY 7
SHIFT FROM EXPLOITATIVE TO REGENERATIVE VALUE CREATION

Current value regimes are predominantly degenerative, they extract value from the natural world in an exploitative value model. Without regeneration, such exploitation will lead to complete degradation of our life-support system. Many initiatives are working on ways to reduce overexploitation of the natural world, for instance by increasing the life spans of products and by upcycling waste streams, think of initiatives such as the re-use centres, the Waste House, Refo, Pass-on shops and repair cafés. Other initiatives are aiming to do even better, instead of reducing negative impact, they are out to create positive impact. We call this 'regenerative value creation'. Initiatives like the Living Coast initiative in Brighton or the transition town and permaculture initiatives in Stockholm, Dresden and Budapest for example, are already taking first steps in this regard by aiming to revitalise communities and the local environment. They are shifting the value logic from degeneration to regeneration. For instance, benefits of urban agriculture initiatives based upon permaculture principles include:

- producing healthy, locally produced food;
- providing meaningful activities for citizens;
- reconnecting citizens to nature;
- promoting healthy exercise;
- building social cohesion;
- restoring the soil;
- sequestering carbon;
- promoting biodiversity;
- regenerating the environment;
- …

while increasing local resilience, fostering wellbeing and encouraging food sovereignty.

Yet, these co-benefits are often not acknowledged, let alone appreciated or rewarded, leaving many of these life-friendly initiatives struggling to survive in today's economic selection environment. Shifting towards a low carbon, life-friendly and equitable society means reconsidering what we value and how. If we are serious about scaling sustainable and meaningful change, then the logic of value creation needs to be changed. As innovation hotspots, cities are ideally placed to lead the way in setting up new value chains and other logics of value creation.

More and more initiatives are giving rise to new value arrangements, just think of citizen energy cooperatives, sharing platforms, alternative currencies, community-supported agriculture initiatives and public-civic partnerships. What is more, transition initiatives actively fuel open source knowledge about value generating processes which in turn supports rapid replication and scaling across the globe.

Critics may claim that these are predominantly symbolic discourses and that implementation is still weak. And indeed, implementation is lagging behind, but that does not mean nothing meaningful is happening. It is just that the current value system, often referred to as value regime in transition literature, is based on old models of thinking (e.g. reductionistic, exploitative and extractive value creation) and thus not designed to incorporate novel ways of arranging value (e.g. multiple, shared, collective and regenerative value creation). This is also a question of powerful vested interests and infrastructure lock-ins that stall system transformation. For example, a fast shift to decentralised renewable energy production is hampered because the current energy infrastructure is centralised and because powerful fossil fuel corporations lobby to delay decarbonisation plans.

Anchoring value-based thinking into policy programs and governance agreements, is the ideal groundwork for inducing new ways of thinking and developing new transaction models that are more appropriate to today's world. But it will only succeed if such values are incorporated in the heart of the city-making processes and are backed-up by new ways of arranging democracy, novel capabilities and transformative leadership strong enough to withstand the pressure of vested interests that are defending the status quo. Cities are already experimenting with incorporating value-based thinking into governance structures: Genk's current policy programs are value-oriented, Brighton incorporated the One Planet Living framework into their governance approach and the Södertälje municipality in Stockholm serves only organic food that has to comply with criteria such as healthy, organic, locally produced food. Scaling value-centric ways of thinking, doing and organising will further be promoted if it is adopted by the investment sector. The Magnet Community Bank in Budapest already works to establish a value-centric society in Hungary by focusing on community building and sustainability. In this, they are not alone. Triodos Bank in the Netherlands, Spain, Germany, Belgium and the UK has already been working more than two decades on sustainability by only investing in activities that are good for the planet. Doing well by doing good. It can be done.

IDEAS TO EXPLORE:
Community bonds   *New business models*   *Basic income*
Time banks   *Local currencies*

Nature builds from the bottom up and everything starts with good soil. Investing in soil is a crucial aspect of a circular economy.

Jos Martens, Master of Compost Genk

TEN THINGS TO KNOW AND DO TO
FAST FORWARD URBAN SUSTAINABILITY

#8.
Updating urban social software is foremost about advancing collective learning and enabling collective intelligence and inclusive action

STRATEGY 8

BUILD CAPACITY FOR EVOLUTION BY INVESTING IN TRANSITION PROCESSES THAT ARE DEVELOPMENTAL, INCLUSIVE AND TRANSFORMATIVE

There is a saying by William James that goes: "A chain is no stronger than its weakest link, and life is after all a chain." What this highlights is that we are in it together – we are all connected. Building the city of the future therefore is a collective endeavour. The logic is simple. We cannot transform unsustainable cities unless we change who we are, what we think and how we act – together. We need to raise our own understanding of how our thinking reinforces the status quo and need to do this collectively. While the work starts with ourselves, success depends on all of us. We cannot afford to leave people behind, everyone needs to go forward. The huge divide between rich and poor is an incredibly weak link and barrier that prevents the upgrading of collective consciousness, intelligence and inclusive action. Referring back to the Albert Einstein quote that was described earlier in the book: resolving the problems of today requires a new consciousness. This requires collective intelligence, genuine equality and distributed leadership to enable and promote self-organisation.

Many urban change makers are doing exactly that, working towards awakening a new consciousness while investing in collective intelligence. Via awareness campaigns such as 'No chemicals, better health' (Zonder is gezonder; Velt Genk), 'Everyone can do something' (Alla kan göra nagot; Refo Stockholm), 'Love food, hate waste' (Brighton & Hove Food Partnership), the 'carbon conversations' (Brighton Peace and Environment Centre), transition initiatives work to expand the consciousness of the residents. By claiming dialogue space, they promote the exchange of ideas and tap into the realm of collective intelligence. Transition initiatives experiment with a multitude of ways to do so, such as participative dialogues, community building approaches geared towards increased resilience, building new collaborative innovation networks, cross-pollinating with other schools of thought and via experiential learning places such as demo gardens, fab labs and compost schools. Often, these initiatives specifically target deprived or vulnerable groups and many initiatives hold values such as inclusivity, equity and equality central in all their undertakings. Examples include the initiatives Refo in Stockholm, Transition Town Wekerle in Budapest and the Pass-on shop and Eco Communal Gardens (called Together Gardens) in Genk. What is more, the majority of the initiatives studied combine both social and ecological dimensions. They showcase that urban sustainability transitions are not just about producing less waste or more renewable energy. They are as much about providing new services as a community and redistributing power and resources.

To advance collective intelligence and enable inclusive action, urban change makers are pioneering new organisational models that are more apt to deal with the dynamics of change. Flat, horizontal networks where leadership is shared and distributed allows for local attuning and more agile responsiveness via processes of self-organisation and self-control. Shared leadership is about having power 'with' instead of power 'over' and distributed leadership is focused on reciprocating rather than controlling activities[18]. These new models give rise to a global network of local responses, termed cosmopolitan localism[19], and thus scale via decentralisation instead of growing ever larger. It is about redistributing power and continuously enhancing distributed intelligence for the effective mobilization of skills. This way, urban change makers are updating our urban social software for the next stage of civilization.

DISCIPLINES TO INVOLVE:

Regenerative design and development

Transition science
Systems thinking

Participatory design
Design thinking

> Fun and laughter glues people together and makes it more interesting to do the things that need to be done. And it is important to know that having fun does not mean that you are not good at your job, it means that you value your time too much to do a boring job.
>
> Liselotte Noren | Refo Stockholm

TEN THINGS TO KNOW AND DO TO
FAST FORWARD URBAN SUSTAINABILITY

#9.
Space is an important resource for innovation. Open up space amid everyday busy-ness in support of business-as-UNusual

STRATEGY 9
CREATE AND INVEST IN ENABLING INNOVATION SPACE (TEMPORAL AND SPATIAL) WHERE IT IS SAFE TO FAIL FORWARD AND WHERE NEW COLLABORATIVE INNOVATION NETWORKS CAN HATCH

Scaling urban sustainability requires the creation of enabling innovation space: space as the incubator for co-creation, recombination, partnership building and disruptive innovation and experimentation. Whether organised citizens reclaim land for nature, initiatives claim dialogue space or spaces for the reparation of stuff or community building space, public initiatives set up new spaces for climate action or businesses establish maker spaces such as FabLabs, BioLabs, Hackathons... they all designate and reclaim space in new ways. Mental space for thought experiments, temporal space for building a shared city vocation, social space for co-creating collectives or cooperatives or physical space for renaturing, experimenting and prototyping new solutions – innovation requires free space. Free space that allows business-beyond-the-usual. Because today not taking risks is the real risk. Mashing up ideas and experimenting with them is how we learn, shift perspectives and discover new solutions. Building collaborative innovation networks is how we bring new solutions into practice. In our highly organised world dominated by efficiency thinking, there is hardly any space to fail forward. We have imprisoned ourselves in the past and try to contain evolution by resisting change. Via their transition initiatives, urban change makers claim space to break out of this confinement. Instead of preventing change, they use change positively to create more change from a learning-by-doing mindset. So, innovation is not only coming from the business world. Our findings show that we can all be creators of a new future. Change makers are the pioneers and transition initiatives the start-ups of a new urban future.

Creating enabling and experimental space is essential for bringing prospective partners together. Partnerships are the first stepping stones of networks and they have been quintessential in the transition initiatives under study. They are critical for unlocking new forms of value, giving rise to new value chains or upgrading value chains into value networks. They also help spread sustainable innovation into new networks and promote the tuning of sustainability solutions to the place-specific context. Unlocking untapped value and place-based development of sustainability solutions therefore requires spaces for co-creation, dialogue and recombination where actors can connect freely and align around compelling shared purposes. Such spaces need to be open, transparent, self-organised and facilitated by experienced mediators, translators and network architects. Ideally, these spaces promote creativity and support disruptive innovation and experimentation, realised by communities of practice and oriented at a shared purpose and collective vision while being open to a plurality of approaches. While we cannot create more space on our planet and in our cities, we can reclaim and utilize space in all its dimensions.

THINGS TO SET UP:

Hackathons	*Fablabs*	*Rule free zones*
Maker spaces	*Living labs*	

TEN THINGS TO KNOW AND DO TO
FAST FORWARD URBAN SUSTAINABILITY

#10.
Forget certainty, stability and control. The future is made at the sweet spot between order & chaos and unity & diversity

STRATEGY 10
INVITE, HOST AND EMBRACE DIVERSITY, IMPROVISATION AND CONTRADICTION

A recurring pattern in all five city regions is that change makers have to overcome many barriers, such as risk-averse mentalities, excessive bureaucracy, silo thinking, short-term thinking, stifling uniformity and sustainability illiteracy to sustain and grow their initiative. These barriers are associated with outdated management paradigms that strive for predictability, stability and control and which have shaped rigid, over-organised urban structures that are often too complex to navigate and too rigid for incorporating novelties. Several times, change makers reported that the siloed structure of the local government proved to be a huge hindrance for developing more holistically-oriented initiatives because these touch upon the jurisdiction of multiple departments each with different sets of rules and ambitions. At the same time, short-termism of local authorities, meaning that support is often provided on a project base, has been identified by the majority of inititiatives as an impediment for evolving and scaling sustainability locally. In addition, the tendency for order frequently pressurizes initiatives in a 'fit and conform' strategy, watering down their capacity for breakthrough innovations. The science however is indisputable: the city as we know it today is terminal. The only way out is to evolve our cities. This requires both chaos and diversity as essential ingredients for building capacity for evolution because they allow recombining the old into something new. Our findings show that transition initiatives contribute to a great extent in promoting diversity locally. And a new role for local governments is to facilitate the right amount of chaos for new urban innovations to unfold while bringing in the right amount of order at the right time to unite novelties into inclusive, evolutionary and regenerative urban innovation ecosystems.

Developing radically new solutions will always be amateurish, imperfect and chaotic at the beginning, and failure is part of the game. The only real failure is the one not learnt from. Yet failure, imperfection and chaos are usually far out of our comfort zone. The converging crises and increasing volatility however, urge cities to start managing for dynamism instead of managing for control by building capacity for evolution into the urban fabric. More specifically, building capacity for co-evolution of human and natural systems so that the viability and vitality of both is supported as the world changes around them. Future success will thus depend on the competence for fast and powerful collective learning. It will require cities to plan for disruption instead of stability. It will require cities to shift from managing the parts to managing the energy flows within the urban ecosystem, allowing order and unity to co-exist with chaos and diversity. In other words, urban innovation will have to combine properties of design with properties of emergence. This means shifting from a command and control to a more improvisational or facilitative management style and shifting from a dominator to a partnership mode focusing on empowerment and emancipation. The growing movement of resilience planning in cities across the world is a good example of dealing with uncertainty while safeguarding diversity. Temporary rule free zones on the other hand, are a good strategy to promote a little bit of chaos to catalyse innovation.

BOOKS TO READ

The Chaos Imperative: How Chance and Disruption Increase Innovation, Effectiveness, and Success | **Ori Brafman**

Reinventing Organizations | **Frédéric Laloux**

The Day After Tomorrow | **Peter Hinssen**

Change of Era | **Jan Rotmans**

Future-Fit | **Giles Hutchins**

Chapter 8:
Conclusion

We give a lot of airtime to building disruptive products and services, to buying and/or investing in disruptive companies, and we should. Both are vital engines of economic growth. But, the most overlooked engine of growth is the individual. If you are really looking to move the world forward, begin by innovating on the inside, and disrupt yourself.

Whitney Johnson, writer

Reinventing ourselves as human beings

Our findings show that change to fast forward urban sustainability is happening organically, everywhere, all the time. The innovators are you and me – by evolving our ways of thinking, our ways of being and our ways of relating to the world we are transforming our responses on a personal and a collective level. Our insights demonstrate that creating the city of the future is most of all a developmental process. It is about Re-thinking, Re-connecting, Re-purposing, Re-naturing, Re-skilling, Re-distributing and Re-organising. Accelerating urban sustainability is therefore also about slowing down. Slowing down to reconnect with our inner and outer nature. Slowing down to recognize our interdependence with the wider community of life. Taking time-out of everyday busy-ness to ask deeper questions: How can we change the role of humans so that we become good for the planet? Sustainability is much more than devising solutions to problems. At its core, sustainability is about making beneficial contributions to the future of life on the planet. It is about making meaningful contributions to our community and our environment. It is about evolving our capacity for co-evolution. And while many of the world's largest private and public institutions fail to move beyond 'management of unsustainablity', devising eco-efficiency discourses focusing on doing less bad (using less energy or producing less waste), treating the symptom rather than the causes, urban transition initiatives do the exact opposite. They get to the root of the problems and work to transform the mindsets that kept the unsustainable system in place in the first place. Many of them are redesigning the way they work and live to generate a positive impact, to regenerate communities and ecosystems. Urban transition initiatives are reinventing the way cities inhabit this planet by fuelling processes to rethink what we value, what is of value and how citizens can shift from value extracting to value adding roles. Our research also shows that upscaling value adding roles depends on new ways of collaboration. This inevitably entails reinventing ourselves as human beings: what kind of human do we need to be so that we can co-create the conditions for beneficial change? How can we become more mindful and build capacity for evolution into our design, organisation and decision-making processes? Building the city of the future is about a whole lot of little things that are locally attuned, meaningful and create positive impact. Most of all, it is about renaturing human nature. Because our nature shapes the nature of cities.

While change is constant, it is never easy

The goal of this book was to showcase how urban change maker initiatives are scaling beneficial and meaningful change to fast forward urban sustainability. By reading about their success stories, you might think that transforming our ways of thinking, doing and organising is easy. The opposite is true. Change is never easy. Change always comes with some form of resistance and we are naturally inclined to prefer pathways of least resistance. Yet, there is no such thing as take-off without resistance, just think of a bird, a bike or a plane. So during our research, we also discovered many examples of stagnation and even deceleration of urban sustainability transitions. But that was not the focus of this book.

The tough conditions for invoking change highlight even more how influential the work of urban change makers really is. Change makers have to be very persistent, tireless and creative to overcome the many barriers they are confronted with. And these barriers are not only induced by the dominant neo-liberal market logic that aims to grow infinitely on a finite planet. Many times they are also the result of antiquated and fossilized governance structures and mindsets and even myopic, short-sighted, reductive scientific approaches. Only strong, determent, adamant and untiring characters can keep going even if the winds blow them so many times off track. The recent austerity discourses make it even more difficult for transition initiatives and too often, initiatives fail to survive long enough to really scale a benevolent impact.

However, transition initiatives play a crucial role in shifting the whole system to a more desirable and sustainable track – even the ones that fail to survive. Just like pioneers in nature, change makers are able to withstand harsh conditions, to mobilise new resources and stabilize the soil for others to come in. No wonder that such pioneering transitioneurs were found to be rare in each of our study sites. They need to be fostered, nurtured and empowered by their cities. To really move forward, cities need to enable disruptors and outliers instead of weeding them out. And once cities start to invest in enabling and empowering conditions, many more still dormant change makers might wake up. In many ways, the five city regions are progressing on this track. Given the urgency and scale of the grand challenges of our times however, more speed is needed.

We have to accelerate urban transitions to sustainability. The only way forward is by becoming involved ourselves and by exponentially growing positive impact. This requires profound collaboration. Time for us to evolve into supercooperators[20]. In this book, we highlight a few essential elements for scaling and speeding up positive sustainable change in cities. It is up to you – the change maker – to devise and build your own locally attuned solution out of these (or others, we do not claim to be exhaustive). For you are better placed than us to know which solution is urgently needed and fits your city.

Together, we can change the world, city by city.
This is no utopia. It is already happening.
Key is to find it, support it, scale it and facilitate it.
The time is now. You are the one you were waiting for.

Endings

– a personal outro

Even though sustainability has been the focal point of this book, our quest for sustainability is only the beginning. There is a big difference between surviving and thriving, between enduring and flourishing. The key, I think, is regeneration, can we redesign our relationship with nature to increase the health, vitality and viability of both? Can we add more value than we extract? I think we can. But we will have to break out of the reductionist box that we created. We have to let go of our desire for predictability and control. And we have to stop thinking that our actions are too small to matter. We all have a value adding role to play. Since I started the book with a story, it seems appropriate to end with one as well.

> There is a little girl that lives in a big city. A city that houses hundreds of millions of creatures, big and small, human and non-human. Walking to school is one of the little girl's favourite things, even though it takes her more than an hour. It is hard to imagine that streets once belonged to toxic exhaustion machines. Nowadays they are meadows blooming with flowers, buzzing with bees, humming with birds and painted with butterflies. She likes the clean air, full of healing aromas produced by the trees that cover the buildings. She loves crossing the bridge over the river, because from there she can spot the bike highway that generates part of the city's solar energy. If she is lucky she might even spot her father, who left early for work. He is a bike doctor and thus an important actor in the city's commuter community. She greets the kingfisher resting on the railing and spots the little fish he has his eye on in the clear water. She strolls on, soaking up the soothing vibrancy of urban nature and enters the food forest, one of her favourite parts of the city. Because this is the best place to spot a badger, a deer, a bear or even a wolf. It deeply grieved her when she learnt in school that wolves were almost hunted to extinction once. Now they are the managers of the food forests, because they keep the systems healthy. A big part of the city's food supply is thus in the hands, or better paws, of the wolves. She stops to pick a fresh blueberry from a bush and is careful not to step onto a line of busy ants. Because also the tiny creatures in this city are treated with respect. Finally, she reaches the school, located in the middle of the

urban food forest. Her lungs full of fresh air, her senses sharp, her brain soaked in oxygen, her being calm yet energized, she is ready for a day of learning. In this school it is not only humans that teach. On the program today is to learn from the whales how to regulate the climate. And even though her city is far from the ocean, the little girl will be diving with the whales within minutes – in virtual reality. Because her city is high on nature and technology.

So my vision for the future are regenerative cities. Cities that teem with all sorts of life. Cities that are beautiful, healthy, prosperous, fair, inclusive, peaceful, abundant and diverse. Cities that function like forests and produce all the ecosystem services life depends upon. Cities that breed creativity, friendship, compassion, love and life. Cities where technology is used to do well by doing good, to create conditions conducive to life. To get there, implementing sustainability in cities is essential but not enough. Let's move beyond aiming for *zero* impact. Let's aim for *positive* impact.

It is not easy waking up from the world we have created. It is not easy to stay motivated when life around us is crumbling down. Yet the choice before each and every one of us is easy. Are you part of the solution or the problem? Are you investing in the future or in the past? Will we entertain ourselves into extinction by staying on the course of business-as-usual? Or will we free ourselves from paralysis, like the hummingbird, to stop agonizing and start organising?

Many change makers are already doing this.
Are you?

Saying thanks

This book would not have been possible without the help of so many people that donated their time, their pictures and their insights. So first off, we would like to thank all the people of the five city regions who cooperated on this study, especially those involved in the transition initiatives, the citizens, helicopter viewers, community workers, civil servants, entrepreneurs, bloggers, scientists, artists, designers and politicians for their openness, interest and time. We truly appreciate your help in co-creating and co-reflecting on how we can speed up the process of sustainability transitions and we learnt a lot from your hands-on experience. A special word of appreciation also to the multidisciplinary research team that was involved in the conception and the execution of this transformative project. Even though we came from different backgrounds, expertise and contexts, which fuelled many thought-provoking, reflective and reflexive debates, collectively we elevated our thinking, moved beyond the silos and explored the boundaries between conventional scientific rigour and innovation-beyond-the-usual. We would also like to thank Ugo Guarnacci from the European Commission for his constant interest and support throughout the project. For both authors of the book, this has been the most collaborative, co-productive and fun project we have ever been involved in. Special thanks also to VITO Belgium's Research Director Walter Eevers who made available funds for the writing of the first version of the book after the project was finished and all the VITO colleagues, especially Kristin Geboers who helped with the practical organisation of making a book. We are very grateful to the LannooCampus team, especially Niels Janssens and Tine Poesen for their support to make this unconventional book happen and to Fulya Toper for the graphic design. Finally, a big thank you to friends and family for their continuous support and to many pioneers and change makers that continuously stretch and elevate our thinking.

This book was made possible by the extraordinary collaboration of a multitude of people

Co-authors of the book:
Jake Barnes (SPRU), Andreas Blum (IOER), Sara Borgstrom (SRC), Markus Egermann (IOER), Niki Frantzeskaki (DRIFT), Ania Rok (ICLEI) and Logan Strenchock (CEU)

Contributors to the research project:
My Almqvist (SRC), Matthew Bach (DRIFT), Jake Barnes (SPRU), Eva van Baren (RSM), Andreas Blum (IOER), Sara Borgstrom (SRC), Sarah Czunyi (CEU), Rachael Durrant (SPRU), Markus Egermann (IOER), Franziska Ehnert (IOER), Thomas Elmqvist (SRC), Anna Emmelin (SRC), Isabel Fernandez de la Fuente (ICLEI), Niki Frantzeskaki (DRIFT), Gyula Gábor Tóth (BEE), Ernst Gebetsroither (AIT), Leen Gorissen (VITO), Clara Grimes (ICLEI), Linda Juhász-Horváth (CEU), Florian Kern (SPRU), Steven Kennedy (RSM), Stefan Kuhn (ICLEI), Derk Loorbach (DRIFT), Gordon MacKerron (SPRU), Steffen Maschmeyer (DRIFT), Karin Markvica (AIT), Michael Meaney (LU), Erika Meynaerts (VITO), Nikolina Oreskovic (SRC), László Pintér (CEU), Kristin Reiß (IOER), Ania Rok (ICLEI), Giorgia Silvestri (DRIFT), Felix Spira (DRIFT), Maria Schewenius (SRC), Dorukhan Sergin (BOUN), Logan Strenchock (CEU), Peter Ulrich (ICLEI), Pieter Valkering (VITO), Allison Wildman (ICLEI), Gail Whiteman (LU), Marc Wolfram (YONSEI UNIVERSITY/ DRIFT), Gönenç Yücel (BOUN)

Dutch Research Institute For Transitions (DRIFT), Erasmus University Rotterdam, the Netherlands
Leibniz Institute of Ecological Urban and Regional Development (IOER), Germany
Science Policy Research Unit (SPRU), University of Sussex, United Kingdom
Flemish Institute for Technological Research nv (VITO), Belgium
Industrial Engineering Department, Boğaziçi University (BOUN), Turkey
Stockholm Resilience Centre (SRC), Sweden
Austrian Institute of Technology (AIT), Austria
Local Governments for Sustainability (ICLEI), Germany
Central European University (CEU), Hungary
BEE Environmental Communication (BEE), Hungary
Lancaster University (LU), United Kingdom

About the authors

Leen Gorissen, PhD is an Innovation Biologist and Sustainability Transitions Expert. She is the founder of Studio Transitio and collaborator of nexxworks and regenerators. Previously, Transition Research Coordinator at the Flemish Institute for Technological Research (VITO), Leen supported participatory action research on topics such as low carbon futures, a circular economy, sustainable forest management, business model innovation and urban transitions. She was the principal coordinator and researcher of the ARTS project in the city of Genk.

Erika Meynaerts is a Commercial Engineer and Researcher at Energyville, a collaboration between the Flemish research partners KU Leuven, VITO, imec and UHasselt in the field of sustainable energy and intelligent energy systems. Erika supports policy makers at different governmental levels in drafting roadmaps to a low carbon future. She was part of the ARTS research team involved in the city of Genk.

Endnotes

1. A combination of the insights of Fred Kofman. 1992. Double-loop Accounting: a Language for the Learning Organisation. *The Systems Thinker 3*, no 1 and Mark Turner. 1996. *The Literary Mind: The Origins of Thought and Language.* Oxford University Press, Inc.
2. This parable originates with the Quechuan people of South America and has been transcribed by Michael Nicoll Yahgulanaas. 2008. *Flight of the Hummingbird: A Parable for the Environment.* Greystone books. This is my recounting of the tale.
3. Inspired by the work of Pamela Mang & Ben Haggard. 2016. *Regenerative Design and Development. A Framework for Evolving Sustainability.* Wiley.
4. The ARTS project (2014-2016) was funded by the European Union's 7th Framework Programme for research, technological development and demonstration under the grant agreement no 603654.
5. These dynamics of change differ from the earlier phases of sustainability transitions, which focus on creating and nurturing innovations.
6. Building on the literature on sustainability transitions and urban governance.
7. An excellent framework for evolving sustainability is described in *Regenerative Design and Development: A Framework for Evolving Sustainability*, Pamela Mang & Ben Haggard, 2016, Wiley.
8. A groundbreaking work on innovating innovation to create conditions conducive to life is described in *Biomimicry - Innovation inspired by nature*, Janine Benyus, 1997, HarperCollinsPublishers.
9. A good overview of the origins of the Wekerle Transition Town initiatives have been described in a report by the TRANSIT project: http://www.transitsocialinnovation.eu
10. ECOLISE, a European network for community led initiatives on climate change and sustainability: A community-led transition in Europe – Local action towards a sustainable, resilient, low carbon future, 2017.
11. ECOLISE, a European network for community led initiatives on climate change and sustainability: A community-led transition in Europe – Local action towards a sustainable, resilient, low carbon future, 2017.
12. Paul Hawken. 2017. *Drawdown: The Most Comprehensive Plan Ever Proposed to Reverse Global Warming.* Penguin Random House LLC.

13. Inspired by Mang & Haggard. 2016. *Regenerative Design and Development: A Framework for Evolving Sustainability*. Wiley.
14. See Giles Hutchins. 2012. *The nature of business. Redesigning for resilience*. Green Books Ltd.
15. Paul Hawken. 2017. *Drawdown: The Most Comprehensive Plan Ever Proposed to Reverse Global Warming*. Penguin Random House LLC.
16. Martin Nowak and Roger Highfield. 2011. *SuperCooperators: Altruism, Evolution, and Why We Need Each Other to Succeed*. Free Press.
17. Adrian Smith and Rob Raven. 2012. What is protective space? Reconsidering niches in transitions to sustainability. *Research Policy*, 41 (6).
18. Inspired by Giles Hutchins. 2016. *Future-Fit*. The Write Factor, UK.
19. Wolfgang Sachs. 2010. *The Development Dictionary: A Guide to Knowledge as Power*. Zed books, London.
20. Martin Nowak and Roger Highfield. 2011. *SuperCooperators: Altruism, Evolution and Why We Need Each Other to Succeed*. Free Press.

Glossary

A* energy-efficient rated building	A building that is the most efficient in terms of likely fuel costs and carbon dioxide emissions according to the UK Energy Performance Certificate system.
Anthropocene	A proposed new epoch that represents human impact on the earth's geology and ecosystems.
Bike train	'Bike trains' are similar to camel trains in that they feature groups of people cycling en masse or as a group. In some cases bike trains move as if they are a single vehicle.
Biodiversity	Or 'biological diversity' generally refers to the variety and variability of life on Earth encompassing variation at the genetic, the species and the ecosystem level.
Biodiversity loss	The ongoing extinction of species worldwide.
Bio-inspired innovation	Innovation inspired by the living world, see also biomimicry.
Biomimicry	The art and science of studying the natural world, translating nature's time-tested and life-friendly strategies and principles and applying these to human design and organisational challenges.
Biophilic design	The integration of elements of nature into architecture and urban planning projects.
Biosphere	The biosphere encompasses the regions of the surface and atmosphere of the earth that are occupied by living organisms.
Blue growth	A long-term strategy to support sustainable growth in the marine and maritime sectors as a whole.

Carbon footprint	The total set of greenhouse gas emissions caused by an individual, event, organisation, product, community or region expressed as carbon dioxide equivalent.
Change maker	A resourceful person that desires change in the world and makes it happen.
Circular economy	An industrial system in which the potential use of goods and materials is optimized and their elements are returned to the system at the end of their viable life cycles.
Citizen budgeting	A practice known as 'participatory budgeting' (pb) allows citizens to determine how a part of government funds are used.
Citizen cooperative	Or co-op, is an autonomous association of people, united voluntarily, to meet a common economic, social, ecological and/or cultural need or aspiration through a jointly owned and democratically controlled business.
City region	The functional region around a city, defined by the functional interrelations between the city and its hinterland.
Cleantech	A shortened form of 'clean technologies', that minimize environmental impacts by improving energy-efficiency significantly, using resources in a sustainable way etc.
Climate adaptation	A response to climate change that seeks to reduce the vulnerability of social and biological systems to relatively sudden change and offset the effects of global warming.
Climate change	A change in global or regional climate patterns, attributed largely to increased levels of carbon dioxide emissions and resulting in e.g. melting ice shields, rising sea levels, more frequent extreme weather events.

Co-creation	A process of involving all stakeholders in the creation of new ideas.
Community bond	An innovative debit instrument that can only be issued by a nonprofit or charitable organisation, enabling them to attract capital. Investors get financial return as well as social return in the form of community benefit outcomes.
Community café	A gathering place where people from all walks of life can meet and talk over a meal. Customers can decide how much they eat and how much they pay for it, making healthy, sustainable food affordable for everyone.
Communities of practice	A process of social learning that occurs when people, who have a common domain of interest, engage in a process of sharing information and experiences to learn from each other and develop professionally and personally.
Consciousness hacking	An approach to change the human relationship to the world that explores technology as a catalyst for psychological, emotional and spiritual flourishing. These include mindfulness and meditation as new tools for the evolution of individual and collective consciousness.
Co-op, co-operative	See citizen cooperative
Crowd sourcing	Soliciting services and ideas from 'the crowd', usually through social media, to expand and diversify the number of ideas.
Customer centricity	Refers to putting the client central in business models and strategies. Translated to the urban environment, this would mean putting the citizen central in urban planning and decision-making processes.

Eco-efficiency	Aims at minimizing ecological damage while maximizing efficiency by reducing the use of resources such as energy, water and materials.
Eco-house	A house that is designed and built using materials and technologies that reduce its environmental impact and energy needs.
Economic resilience	The ability of an economy to recover from or adapt to the impact of exogenous shocks such as natural disasters, financial crises etc.
Ecosystem services	The direct and indirect contributions of ecosystems (e.g. forest, aquatic, grassland…) to human wellbeing, that can be categorized in provisioning services (e.g. food), regulating services (e.g. climate regulation), supporting services (e.g. pollination), cultural services (e.g. recreation).
Ecosystem stewardship	Or environmental stewardship refers to the responsible use and protection of ecosystems, by means of conservation and sustainable practices, to enhance ecosystem resilience as well as human wellbeing. It recognizes resource managers as an integral part of the systems they manage.
E-mobility	A road transport system based on transport modes that are powered by electricity, such as electric vehicles, electric (motor)bikes.
Empowerment	A method to increase the degree of autonomy and self-determination in people and in communities to enable them to represent their interests in a responsible and self-determined way, acting on their own authority.
Energy efficiency	Also efficient use of energy, aims to reduce the amount of energy used to provide the same services or produce the same product.

Environmental stewardship	See ecosystem stewardship
Environmental sustainability	A dimension of sustainability that focuses on promoting environmental integrity and restoration. It works towards responsible interaction with the environment to avoid depletion or degradation of natural resources and allows for long-term environmental quality, ensuring that the needs of the current population are met without jeopardizing the ability of future generations to meet their needs.
Envisioning	The process of imagining and visualizing the future.
Equitable city	A city that treats its citizens fairly and equally by considering principles of equity, democracy and diversity during the city planning and policymaking processes.
Fab lab	A cooperative workplace where inventors and developers can use collective infrastructure, such as computers, 3D printers, for creating new services and/or products.
Food forest	A low maintenance sustainable plant-based food production and agro-forestry system based on woodland ecosystems that produces food for humans and all other forms of life. It is an intermixed growth production system of many layers encompassing nut- and fruit-producing trees, vines and perennials that produce fruits, vegetables, and edible greens and roots.
Forest bathing	Or Shinrin-yoku is a term that means 'taking in the forest atmosphere'. It was developed in Japan during the 1980s and has become a cornerstone of preventive health care and healing in Japanese medicine and beyond. Research is increasingly showcasing the nature therapy benefits for human health, wellbeing and cognitive functioning.

Future-proof, future fit	Refers to a state that is responsive to change, i.e. that has the capacity to adapt, evolve and flourish in times of change and disruption.
Greenhouse gas (GHG)	A gas, e.g. carbon dioxide and chlorofluorocarbons, that contributes to the greenhouse effect by absorbing infrared radiation.
Governance context	All the processes that are involved in governing an organised society whether undertaken by the government, market or civil society. The governance context refers to the processes of interaction and decision-making among all the actors in the city that shape, reinforce and reproduce social norms and institutions.
Governance experimentation	Innovative cross-sector or cross-domain experiments aimed to radically change governance frameworks.
Green infrastructure	Strategically planned network of natural and semi-natural green (land) and blue (water) spaces, designed and managed to deliver a wide range of ecosystem services such as water purification and captation, air quality, recreation, climate mitigation and adaptation.
Hackathon	Gathering of a group of people that engage in creative problem solving, usually lasting for several days.
Impact investment	An investment made with the intention to generate social and environmental impact as well as financial return.
Living lab(oratory)	An innovation platform for knowledge-sharing, collaboration and experimenting in open, real-world environments.
Local currency	Also complementary or community currency, can only be used within a designated geographic area and, often, within a specific time frame.

Low energy building	A building designed to use less energy, including zero-energy building, passive house and green building.
Maker space	A collaborative work space inside a school, library or separate public/private facility for making, learning, exploring and sharing that uses high tech to no tech tools. They are also called fablabs when they offer maker equipment like 3D printers, laser cutters etc.
Mindfulness	A state of paying attention on purpose, in the present moment, which is open, compassionate, non-judgmental and curious. It promotes a more harmonious way of living in the moment and is conducive for greater wellbeing and perceived health.
Multi-level governance	Refers to the different levels that are involved in governing a city: municipality, city, region, nation and international bodies. Responsibility and authority is generally shared across multiple levels which influence autonomy.
Nature-based solution	Solutions that are inspired and supported by nature and that simultaneously provide environmental, social and economic benefits while building resilience. Such solutions bring more, and more diverse, nature and natural features and processes into cities, landscapes and seascapes, through locally adapted, resource-efficient and systemic interventions.
Open innovation	Refers to distributed innovation processes based on purposeful cross-pollination of ideas and knowledge across sectors and disciplines.

Organic food

Food grown by methods that comply with the standards of organic farming such as cycling resources, promoting ecological balance, conserving biodiversity while constricting the use of certain pesticides and fertilizers. Organic foods are usually not processed using irradiation, industrial solvents or synthetic food enhancers.

Participatory budgeting

A practice that allows citizens to determine how parts of government funds are used. See also citizen budgeting.

Pattern thinking

A way of understanding reality by focusing on the relationships between parts of a system rather than on the parts themselves. Pattern thinking is part of systems thinking.

Permaculture

A contraction from the words 'permanent agriculture' and 'permanent culture' that represents a form of living systems design encompassing social and agricultural systems. It is a design philosophy and practice that emulates principles from the living world to create human (production) systems that function in harmony with nature.

Positive psychology

A new domain in psychology that instead of focusing on treatment of 'mental illness', focuses on promoting positive human functioning and flourishing on multiple levels that include the biological, personal, relational, professional, institutional and cultural dimensions of life.

Public-civic partnership

An official partnership between the public sector and civil society where civil servants and volunteers co-invest resources and time to create reciprocal social or ecological benefits.

Rain garden

A depressed area in the landscape that collects rain water from a roof, driveway or street and allows it to soak into the ground.

Reductive thinking	The tendency to present a subject or problem in a simplified form and the belief that by understanding the parts, one can understand the whole.
Regenerative city	A city that adds more value than it takes from the earth by increasing it's capacity of positive co-evolution between human and natural systems.
Regenerative design	A design philosophy and practice that aims to add value to natural systems by supporting processes of co-evolution between human and natural systems, shaping human systems that are conducive to life.
Regenerative development	A development philosophy and practice that aims to regenerate the living world building on the perspective that humans have a value adding role to play within nature. It shifts from designing things to designing 'capability' to support the positive co-evolution of humans and natural systems.
Regenerative leadership	A new leadership paradigm that strives for a conscious way of leading, living and thriving by developing life-affirming cultures geared for the current and future challenges of our times. Regenerative leadership aims to create future-fit organizations where purpose, people, planet and profit can thrive collectively.
Resilient – Resilience	The ability of a system to recover from perturbations and to adapt well to change.
Rule free zone	A novel strategy to promote innovation by taking away overly complex, outdated or hindering rules during a period of time to catalyse innovation beyond the usual.
Service-based leadership	A leadership style that prioritizes the team's needs above self interest.

Smart city	A city that incorporates information and communication technologies to enhance the quality and performance of urban services, reduce resource consumption, wastage and overall costs.
Social cohesion	The extent to which inequalities, exclusions and disparities, based on ethnicity, gender, class, nationality, age, disability or any other distinctions, are reduced and/or eliminated in a planned and sustained manner.
Social economy	Often referred to as a third sector among economies between the private (business) and public sectors (government) that ecompasses cooperatives, nonprofit organisations, social enterprises and charities.
Social enterprise	An enterprise that is operating in the social economy and whose main objective is to have a social impact rather than to make a profit for its owners or shareholders.
Social sustainability	A dimension of sustainability that focuses on promoting social equality and cohesion.
Societal system	The complex system of interrelations between human, technological and natural systems that form a whole.
Stakeholder	A person or group of persons that have an interest or concern in the success or failure of something, such as a business or project.
Self-organisation	The ability of a system to structure itself, to create new structures, learn, improve and diversify.
Sustainability	Meeting the needs of today without compromising the ability of future generations to meet their own needs.

Sustainability transition	The process of shifting culture, structure and practice to higher levels of sustainability. Transition theory addresses transformative change through concepts of co-evolution, multi-level and multi-phase dynamics and co-design and learning.
System	A set of elements or parts which are coherently organised and interconnected in a pattern or structure that produces a characteristic set of behaviours, often termed function or purpose.
Systemic solution	Also termed root cause solution: a solution that addresses the root cause of a problem and thereby radically changes the way a system is organised and structured. Systemic solutions are thus more encompassing than solutions that merely treat the symptoms.
Systems change	Emergence of new patterns of organisation or new structures that radically change the behaviour, function or purpose of the system.
Systems thinking, Living systems thinking	A way of understanding reality by focusing on patterns and interrelationships between parts of a system rather than on the parts themselves.
Time bank	An alternative monetary system based on the value of one hour of work; hours are traded by using or providing services that are often focused on community outreach such as elderly care, home repair etc.
Transformative change	A process of fundamental or marked change in the form, nature or appearance in parts of a system.
Transition	A radical change in the structures, cultures and practices of a societal system.

Transition initiative	A locally-based activity which aims to drive transformative change of existing societal systems towards higher levels of environmental sustainability.
Upcycling	Converting low-value materials into high-value products.
Urban change agent	Someone that brings about or facilitates transformative change in a city.
Urban agriculture	Also urban farming or gardening; the practice of cultivating, processing and distributing food in or around a village, town or city.
Urban governance	The process through which the local government and stakeholders in the city (e.g. business associations, unions, citizens) come to a decision about how to plan, finance and manage the city.
Urban planning	Planning and design of urban space and activities.
Vocation	A calling, a source of meaning that provides a sense of direction for the future.

Find out more

- On the ARTS project website, you can find a lot more information about the project, including a tailored roadmap for accelerating the sustainability transition for each city region under study and the viewpoints of local bloggers on sustainability in their cities: http://acceleratingtransitions.eu/
- The ARTS project is also featured on the website of the European Commission: Mapping out routes towards environmentally sustainable cities, see http://ec.europa.eu/research/infocentre/article_en.cfm?&artid=44496 and in a publication of the European Environment Agency: Sustainability transitions - Now for the long term, see: https://www.eea.europa.eu/publications/sustainability-transitions-now-for-the
- Ehnert, F.; Frantzeskaki, N.; Barnes, J.; Borgström, S.; Gorissen, L.; Kern, F.; Strenchock, L.; Egermann, M. 2018. The Acceleration of Urban Sustainability Transitions: a Comparison of Brighton, Budapest, Dresden, Genk, and Stockholm. *Sustainability* 10, 612.
- Ehnert, F; Kern, Fl; Borgström, S; Gorissen L; Maschmeyer, S; Egermann, M. 2017. Urban Sustainability Transitions in a Context of Multi-level Governance: A Comparison of Four European States. *Environmental Innovation and Societal Transitions*. In press.
- Frantzeskaki, N., Borgstrom, S., Gorissen, L., Egermann, M., and Ehnert, F. 2017. Nature-based solutions accelerating urban sustainability transitions in cities, in Kabisch, N., Korn, H., Stadler. J., and Bonn, A., (Eds), *Nature Based Solutions to Climate Change Adaptation in Urban Areas Linkages between Science, Policy and Practice*, Springer, ISBN: 978-3-319-53750-4, DOI 10.1007/978-3-319-56091-5.
- Haase, D., Kabisch, S., Haase, A., Andersson, E., Banzhaf, E., Baro, F., Brenck, M., Fischer, L.K., Frantzeskaki, N., Kabisch, N., Krellenberg, K., Kremer, P., Kronenberg, J., Larondelle, N., Mathey, J., Pauleit, S., Ring, I., Rink, D., Schwarz, N., and Wolf, M. 2017. Greening cities – To be socially inclusive? About the alleged paradox of society and ecology in cities, *Habitat International*, 64, 41-48, http://dx.doi.org/10.1016/j.habitatint.2017.04.005
- Hartz-Karp J, Gorissen L. 2017. The role of transition initiatives in urban sustainability. In: *Methods for Sustainability*. Marinova Eds. Edward Elgar publishers.
- Loorbach, D., Frantzeskaki, N., Avelino, F. 2017. Sustainability Transitions Research: Transforming Science and Practice for Societal Change, *Annual Review of Environment and Resources*, Vol. 42. In Press.

- Valkering P, Yücel G, Gebetsroither-Geringer E, Markvica K, Meynaerts E, Frantzeskaki N. 2017. Accelerating Transition Dynamics in City Regions: A Qualitative Modeling Perspective. *Sustainability*. In press.
- Frantzeskaki, N., Dumitru, A., Anguelovski, I., Avelino, F., Bach, M., Best, B., Binder, C., Barnes, J., Carrus, J., Egermann, M., Haxeltine, A., Moore, M.L., Mira, R.G., Loorbach, D., Uzzell, D., Omman, I., Olsson, P., Silvestri, G., Stedman, R., Wittmayer, J., Durrant, R., and Rauschmeyer, F. 2016. Elucidating the changing roles of civil society in urban sustainability transitions, *Current Opinion in Environmental Sustainability*, 22, 41-50.
- Frantzeskaki, N., Haase, D., Fragkias, M., and Elmqvist, T. 2016. Urban transitions to sustainability and resilience, *Current Opinion in Environmental Sustainability*, 22, iv-viii.
- Gorissen L, Spira F, Meynaerts E, Valkering P, Frantzeskaki N. 2016. Moving towards systemic change? Investigating acceleration dynamics of urban sustainability transitions in the Belgian City of Genk. *Journal of Cleaner Production*. In press.
- Wolfram, M., Frantzeskaki, N., and Maschmeyer, S. 2016. Cities, systems and sustainability: status and perspective of research on urban transformations, *Current Opinion in Environmental Sustainability*, 22, 18-25.